SpringerBriefs in Applied Sciences and Technology

Computational Intelligence

W0080428

Series Editor

Janusz Kacprzyk, Systems Research Institute, Polish Academy of Sciences, Warsaw, Poland

SpringerBriefs in Computational Intelligence are a series of slim high-quality publications encompassing the entire spectrum of Computational Intelligence. Featuring compact volumes of 50 to 125 pages (approximately 20,000-45,000 words), Briefs are shorter than a conventional book but longer than a journal article. Thus Briefs serve as timely, concise tools for students, researchers, and professionals.

S. D. Varwandkar · M. V. Hariharan

Fractal Tomography
for Power Grids

 Springer

S. D. Varwandkar
Department of Electrical Engineering
Veermata Jijabai Technological Institute
Mumbai, Maharashtra, India

M. V. Hariharan
Department of Electrical Engineering
Indian Institute of Technology Bombay
Mumbai, Maharashtra, India

ISSN 2191-530X   ISSN 2191-5318 (electronic)
SpringerBriefs in Applied Sciences and Technology
ISSN 2625-3704   ISSN 2625-3712 (electronic)
SpringerBriefs in Computational Intelligence
ISBN 978-981-99-3442-3   ISBN 978-981-99-3443-0 (eBook)
https://doi.org/10.1007/978-981-99-3443-0

This Springer imprint is published by the registered company Springer Nature Singapore Pte Ltd.
The registered company address is: 152 Beach Road, #21-01/04 Gateway East, Singapore 189721,
Singapore

For
Aditi, Neha, Neil, Nikhil
from S. D. Varwandkar

Sanghamithra, Arnav
from M. V. Hariharan

Preface

The real joy of discovery exists not in seeking new lands, but in seeing with new eyes.

—*Marcel Proust*

This book results from authors' continuing efforts to critically examine prevalent methods of power system analysis and is a sequel to *Modular Load Flow for Restructured Power Systems* published by authors in 2016 (*LNEE* Series 374, Springer 2016). Reliable operation of power systems is getting more and more difficult with renewable sources. Inertia of systems has taken a toll. As a result, system disturbances cause oscillations even in remotely located generators. Loss of these low-inertia generators makes the system more fragile. Frequency control has become difficult. Voltage and frequency ride-throughs are required. Renewables sources cause concern to the system operator. We therefore ask simple but pertinent questions: Is Jacobian-based power flow a sacrosanct way to compute lineflows and voltages with renewable sources? To what extent eigenvalue and mode-decompositions are relevant to system regulators for the physical control actions that they are called upon to take? These methods do not satisfactorily address the basic issues. Having taught the subject of power systems over several years, the authors felt that this is the time to go back to physics and look for new approaches. The premise in this book is that a power grid is like a human body in which lineflows, frequency and cascading failures mimic blood flow, pulse rate and paralytic conditions of human species. Grids suffer trauma-like conditions, too. Blackouts may result when the system is stressed beyond limits. Smaller perturbations can occur occasionally or frequently. A new concept of *fractal tomograph* (FT) is introduced to study such perturbations as also the precarious conditions in power systems. It captures critical features of the system and acts like a CT scan that helps diagnose normal and abnormal conditions. Medical images have strong similarities with informational numeric cross section of circuit elements from which electrical properties of the grid can be derived. Numerics in the tomograph are called *fractals* and indicate influence of generators on lineflows and voltages. A mathematical phrase termed *Active Network Twins* (ANT)—like DNA in a human body—is discovered which is present in all lineflow and voltage expressions. ANT appears as a multiplier with generator powers in computing *fractals* of the tomograph, and thereby the lineflows and voltages. Fractals can be viewed as carriers of MWs and voltages from generators to loads. Sum of appropriate fractals constitutes lineflows and voltages which turn out to be linear functions of turbine powers. The constant of proportionality is a nonlinear function of *circuit parameters* and is constant for a given network. Lineflows and voltages provide a feel to grid operators which they can use to assess health of the system and control it when desired. Outages possess a past, a present and a future. FT can explain what went wrong in the past and how it could have been prevented, and how to plan in future. *What is remarkable is that a minute detail of the humungous tomograph can indicate the chequered past, an ailing present or a catastrophic future of the system!* It also suggests the mechanism for editing fractals to rectify the malaise and prevent its occurrence. *Being able to spot a particular fractal in a large collection of numerics is a testimony to the extraordinary power of fractal tomography.*

Prevalent methods can solve only specific problems, for example, Newton–Raphson solves the power flow, Lagrange-based optimization solves the economic dispatch, distribution factors, the contingency analysis and so on. Contrary to this, all problems of the system along with their solution are embedded in the fractal tomograph. For example, variability of injections, outages of transmission lines and generators, load-shedding, frequency dynamics, voltage/frequency-ride-throughs in low-inertia systems, composition of lineflows and load voltages, and reconstruction of scenarios in the post-event diagnosis of blackouts are all *hidden within the fractals of a tomograph.* Fractal tomographs, which do not need iterative computation, are useful for quick operational decisions.

Phenomena of synchronizing arise naturally in power grids and indicate system's intrinsic ability to balance itself during disturbances. This ability may reduce significantly in stressed conditions and trigger cascading failures. Entire scene strikingly resembles the condition of a patient under intensive medical care. Cascading failures that result in blackouts are like trauma causing human death. Grid operators, to the best of our knowledge, currently do not have effective tools to analyse this situation. Fractal tomographs provide analytical description of the scenario at every stage. Causality and transparency are its hallmarks. Wide-ranging phenomena in power systems can be explained with fractal tomographs.

Tomographic analysis requires some fundamental changes in our outlook, nay, our insight. Since current flows continuously, it must "return" and form a current-loop, causing voltage drops across resistances in its path. The "ground" bus is introduced in power systems to account for this return. The fact is that it is the "power" that travels from generators to the loads. Power does not "return"! Power transfer is an end-to-end process unlike the current which must form a loop. "Square of the current" can explain transfer of power from one location to another more logically. The I-square pulses travel and carry power from generator to loads where they sink. I-square therefore seems to be better choice than the phasor current as variable for power system analysis. The analytical domain then changes from "KCL" to "square domain" (SD). I-square becomes the SD-current. Circular geometry of 2π becomes a linear geometry on its diameter. Generator-MWs can then be sectioned into "power drops" along the diameter when SD-currents cross resistances, be it load or a line. Laws in the square domain are expressed differently than the Kirchhoff's laws.

Frequency turns dynamic during disturbances. Synchronizing phenomena can be viewed as a *rendezvous* effort by various generators to arrive at a consensus frequency. A new concept of characteristic frequencies of generators is introduced. Consensus frequency is weighted average of characteristic frequencies. The composition displays *fractal* property of the consensus. Frequency properties are required for diagnosing system oscillations, detection of coherency, determination of voltage/frequency overrides, etc. Consensus frequency also affects impedances and thereby the tomographs. *Frequency-rendezvous dynamics* proposed here lends itself to computation of dynamic lineflows and load voltages at each frequency step.

Blockchain has made it possible to keep track of financial transactions in oligopolistic money markets. A completely new topic of blockchain for MWs is introduced in the book. MW-transactions can be tracked using tomographs. It can be used by regulators for logical pricing of electricity and proper disbursal of revenue.

The book is divided into seven chapters. Chapter 1 is a simple and illustrative introduction to brief history of fractals. Concept of Active Network Twins (ANT) which is an important ingredient of fractal tomographs is introduced in Chap. 2. ANT and the biological DNA are explained and discussed in this chapter. Expressions for lineflows and voltages in terms of fractals are derived in Chap. 3. Fractals of tomographs can be edited like genes in the DNA. Editing is explained and illustrated on the 39-bus New England system in Chap. 4. Phenomena of loop flows which caused major concern to regulators in USA around 2007 forms the content of Chap. 5. How fractals can precisely identify the cause of loop flows is illustrated in this chapter. Tomographs possess capability to analyse *present* and predict *future* resulting from outages in the grid. The *past* too can be peeped into. Important blackouts containing all the three stages are analysed in Chap. 6. Frequencies of isolated generators make up fractals which in turn compose the *frequency of rendezvous* (system frequency) equation for which is formulated and discussed with examples in Chap. 7. A new blockchain approach is presented to compute user-end costs which is presented in Chap. 8. Fully illustrated examples with complete data (at the cost of frown from some readers) along with intermediate numerical results are provided to enable discerning readers to rework and double-check the results. Fractal tomography should prove useful in planning, operation, control and forensic analysis of power grids. It should also help market regulators to prescribe settlement policies and design new market procedures.

We hope that grid operators, power professionals, researchers and academia will find thought-provoking material in this book.

Mumbai, India   S. D. Varwandkar
                                                                        M. V. Hariharan

Acknowledgments

The authors are thankful to Prof. K. M. Kulkarni for insightful discussions and constant encouragement. Former Member of Maharashtra Electricity Regulatory Commission, Shri V. L. Sonavane, broadened our horizon by sharing his experience and knowledge. We are much thankful to him. Dr. Rajamani of Adani Electricity Mumbai Limited, Shri Girish Jawale of Tata Power Company Limited and Shri Vasant Pande of Maharashtra Electricity Transmission Company Limited helped us with useful information on Mumbai distribution network for which we are thankful to them. Dr. Jeremy Lin, earlier with PJM, shared a couple of little-known details of the US–Canada blackout of August 2003. We are thankful to him for sharpening our understanding. The first author benefited immensely from Shri K. N. Nafde, a steel industry veteran who spent more than four decades in operations and projects of global steel multinationals. We are thankful to him for sharing his experiences on very large systems. Mahima (Rimzim) went through the text and made some useful suggestions for which we are thankful to her. We are thankful to anonymous reviewers of our papers who favoured us by reading our work and providing valuable comments on our submissions. Manjit and Akalpita (V) offered whole-hearted support in our endeavours. Acknowledging them goes beyond words. Gracious Mrs. Malathi (H) always inspired us for which we are grateful to her. We are thankful to Mohana and Murali (family, V), Ramani and Vaishali, Murali and Anjali (family, H) who extended their affectionate support throughout. Some of the ideas in the book were presented to graduate students of Prof. S. R. Wagh and Prof. N. M. Singh of Electrical Engineering Department, VJTI, Mumbai, in the form of informal lectures/discussions. Our sincere thanks to all the students and faculty at VJTI. Special mention must be made of Shri M. S. Fulpagare who wrote some MATLAB routines for us. We are indebted to Varsha and Ajit (Raipur) for loving care of SDV during crucial times.

For long years, S. D. Varwandkar worked as faculty at VJ Technological Institute, Mumbai and M. V. Hariharan* at Indian Institute of Technology, Mumbai. We place on record our gratitude to our institutes. (* Prof. M. V. Hariharan (90) passed away while the book was under print - SDV.)

When the first author had shown his name on the cover page of the earlier *Springer* book to his granddaughter Aditi, the little angel quickly scanned the first few pages and innocently asked "where is *my* name in it?" The wisdom that dawned at that moment was so sudden that it left the author dumbfounded. Aditi, I (S. D. Varwandkar) am happy to mention your name *first* among the others on the dedication page of this book!

Last but not least, we thank *Springer* for considering our work for publication.

S. D. Varwandkar
M. V. Hariharan

Contents

Abbreviations

AEML	Adani Electricity Mumbai Limited
AEP	American Electric Power Company
ANT	Active Network Twins
CT	Computerized tomography
DNA	Deoxyribonucleic acid
FR	Frequency of rendezvous
FRD	Rendezvous dynamics of frequency
FT	Fractal tomograph
G, B	Conductance, susceptence
ICT	Interconnecting transformer
IESO	Independent Electricity System Operator
ITC	International Transmission Company
METC	Michigan Electric Transmission Company
MISO	Midwest Independent System Operator
MSETCL	Maharashtra State Electric Transmission Company Limited
NYISO	New York Independent System Operator
PAR	Phase angle regulator
PJM	Pennsylvania–New Jersey–Maryland Interconnection
R, X	Resistance, reactance
SD	Square domain
TECL	Tata Electric Companies Limited
UEC	User-end costs

Symbols

P_i	Turbine power of generator i
$[y]$	Primitive admittance matrix
y_e	Admittance of element e
A	Node-element incidence matrix, $A_1 - A_2$
Z	Z-parameter matrix of network
ξ_{ei}	$(e-i)^{th}$ element of $A^T Z = [A^T Z]_{ei}$
p_{ei}	Element-power of e due to generator i
p_e	Element-power due to all generators
q_{ei}	Reactive power of element of e due to generator i
I_{ii}^2	Current-square of the generator in single generator module; SD-current
p_{Cf}	Power flow on cutset C
v_e^2, i_e^2	SD-voltage and SD-current of element e
X	$\omega_0 L$
$\dot{X}$	$\dot{\omega} L$
ε_{ei}	Fractal coefficient of element e with respect to generator i
λ_{si}	Characteristic frequency of generator i rad/s
λ_i	Frequency of generator in conceptual one-generator module, rad/s
ω_i	Actual frequency of generator i, rad/s
ω_s	Frequency of rendezvous of all generators, rad/s
k_i	Frequency-fractal coefficient for generator i
p_{Ldg}	Power of loads connected at generator buses
p_{Ldx}	Power of loads connected at non-generator buses
C_{Ldg}	MW cost of loads at generator buses, vector
C_{Ldx}	MW cost of loads at non-generator buses, vector
p_f	MW-flows on transmission lines, vector
$C_{\left(\frac{v^2}{v_{base}^2}-1\right)}$	Cost of voltage deviations, vector
C_{LL}	Cost of line losses, vector
C_{fi}	Cost of MW-flow on line i

List of Figures

List of Tables

Chapter 1
Introduction

1.1 Brief History

Similarities observed in trees, leaves, flowers, creatures, etc. have enticed researchers for long. Mandelbrot made formal efforts in 1960s and 70s [1] to mathematically represent geometric similarities. A complex number was squared and another added to it. The computation was repeated recursively many times. The plots of numbers on the complex plane started showing repeated patterns after a large number of iterations. These plots were called Mandelbrot sets. Julia sets are similar but result from computation with the number that is added at each iteration. Plots of some natural fractals, Mandelbrot set and Julia set are shown in Figs. 1.1 and 1.2 which which are reproduced from [2] for illustration.

Disclaimer: No copyright infringement is intended.

1.2 Hausdorff Numbers

Mathematical description of shapes gives rise to Hausdorff dimensions [2] wherefrom *numbers* get associated with the *shapes* they represent. For example, Hausdorff dimension for a point is 0, for line is 1, for square is 3 and for cube it is 4. This assignment is purely subjective. There appears to be no physical or causal way for one-step *connect* of patterns with the numbers or vice versa. The excellent book [3] "Fractals in Engineering Systems" illustrates many examples and compilation of mathematical works by researchers. In spite of rich mathematics and mind-blowing patterns, the *mutual physical dependence* that the natural fractals display continues to elude researchers. Nature has both, a spatial and a temporal spread which defines its workspace. Leaves of a plant are similar but seldom identical. Internal structure of leaves is also similar but not identical. Dissymmetry is thus embedded in the natural similarity. Similarities of buds and fruit are temporal spread. Parts of

S. D. Varwandkar and M. V. Hariharan, *Fractal Tomography for Power Grids*, SpringerBriefs in Computational Intelligence, https://doi.org/10.1007/978-981-99-3443-0_1

Fig. 1.1   Fractals in nature

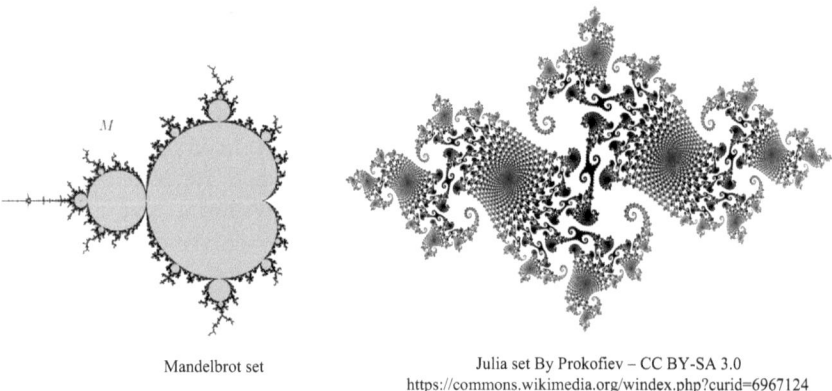

Mandelbrot set   Julia set By Prokofiev – CC BY-SA 3.0
                                          https://commons.wikimedia.org/windex.php?curid=6967124

Fig. 1.2   Fractal plots for Mandelbrot and Julia sets

a plant are physically connected and influenced by one another in some unknown way. However, nature does not produce perfect symmetry. It embeds dissimilarity, too. Perfect symmetry goes against the principles of natural similarity. R. E. Kalman had said, "Get the physics right, rest is mathematics". This statement has been the guiding light in our pursuit for fractals in power grids. We are not carried away by the "beauty" of symmetric and the awesome mathematical plots. We explore the genesis—the "seed" or the "DNA" of the similarity. Biological sciences do not lend easily to mathematical analysis, the power grids do. Grids have topological structure like plants and humans. Power grid has a control centre like brain, network and loads like body parts, and so on. The dynamic behaviour arises from exchange of energies. Power grid works on the principles of power balance and Kirchhoff's laws. Therefore, it is an appropriate area for exploring fractal math. Absence of causality has been instrumental in restricting application of Mandelbrot theory to biological sciences or electrical networks. We incorporate causality in fractal math.

1.3 Fractals in Power Grid

A subtle difference between Mandelbrot approach and the proposed approach is that physical *topology* does not play any role in Mandelbrot/Julia sets, but it *does* in MW/voltage patterns in power systems. Hausdorff dimension produces symmetric patterns after hundreds of recursive computations which severs its connection with the original number. Plant life admits growth and so do power systems. Electric power grids are energized by generators—its *genes*. Power balance principle and Kirchhoff's laws allow us to write equations that lead to fractal expressions for MW/voltages which lend themselves to tomographic analysis. Connection with the "seed" is not severed but is always present in the fractals of tomographs. Essential difference between numerical and tomographic analysis of fractals is summarized below.

Fractal property	Numerical approach	Tomographic approach
Model	Complex plane (Purely mathematical)	Real line scalars (Physical)
Analysis	Based on pure mathematics	Based on physical laws
Patterns	Artificially symmetric	Similar, not symmetric
Validation examples	Simulation plots	Causal relations
Topology	Not considered	Considered
Numbers	Geometric patterns are denoted by Hausdorff numbers	Numbers generated from fractal expressions denote MW and voltage patterns

Power grids generate their own patterns of MW/voltages in the power network. We discover the causality relation that generates these patterns. Electrical circuits with a *scalar power source* are used (*It may be noted that books on circuit analysis deal only with voltage and current—phasor sources.*). A few manipulations yield expressions for fractals which can be used to solve many problems of the power grid, especially, prediction of contingencies and their after-effects, preventive control of anticipated abnormalities, management of congestion or flow-reversals, etc. These are important considerations for the grid operator. In tomographic approach, established physical laws are employed to construct functions which produce Hausdorff-like numbers for MW/voltage patterns of the individual generators. The analysis thus shifts from mathematics to physics. Causality principle offers the advantage that the fractals can be used by system operators for forward computation to control the grid or backward computation to investigate its past as in blackouts. Topology, node-element matrix, and the circuit parameters play pivotal role in derivations. Ensemble of fractals is termed as *tomograph*. There are lineflow tomographs and voltage tomographs which act like CT-scans, and which can discover abnormalities such as, *excessive lineflows, over-voltages/frequency or flow-reversals,* etc. As an illustration, consider resistances of 2 and 98 Ω in series fed by a 100 W turbine-generator source (units chosen to simplify understanding). Power balance principle dictates that 2 W be consumed by

2 Ω resistance and 98 W, by the 98 Ω. The ratios of these with respect to the injected power are 2/100 and 98/100, respectively. These ratios are obtainable *from only the network specifications*. Following workout is self-explanatory.

Series circuit:

$$I^2 = \frac{P}{(2 + 98)} = 10$$

Voltage-squared across 1 Ω $= (I^2 R) R = 10 \times 2^2 = $ constant 1 $* P$.

Voltage-squared across 98 Ω $= 10 \times 99^2 = $ constant 2 $* P$.

Parallel circuit:

$$R = \frac{2 * 98}{2 + 98}$$

Common $I^2 = \frac{P}{\left(\frac{2*98}{2+98}\right)} = \frac{100*P}{98*2}$.

$$V^2 = I^2 R^2 = \left(\frac{100 * P}{98 * 2}\right) * \left(\frac{2 * 98}{2 + 98}\right)^2$$

Current-squared in 2 Ω $= $ constant 3 $* P$

Current-squared in 98 Ω $= $ constant 4 $* P$

Common voltage-squared $= $ constant 5 $* P$.

Values of constants on the RHS represent voltage/current (-squared) patterns of the two power circuits. Note that power P is proportional to I^2, the phasor $\bar{I}$ being a complex number. Exponent '2' can be interpreted as the Hausdorff dimension since one can write $I^2 = a^2 + b^2$ which represents a triangular or semi-circular geometry. The Pythagorean expression is the algebra for circular geometry. Physics of electric circuits thus gets associated with both its geometry and algebra. Arcs on the circular periphery can be projected on the diameter to form linear segments—the *fractals*. Analytical expressions for fractals are derived in the next chapter. MWs/ voltages being scaled values of the generator injection are therefore its linear *fractals*. Interestingly, a mathematical phrase, termed as *Active Network Twins* (ANT), is found in all MW/voltage expressions. It resides throughout the network like its DNA. The ANT *apropos* the DNA is defined and explained in the next chapter. An ANT is specific to a load-generator, or line-generator *pair* and constitutes similarity at a particular topological location.

Fractals assembled in a matrix forms a *fractal tomograph*. Like a CT scan, it contains valuable information about the system. This book demonstrates the use of fractal tomographs for several day-to-day operational problems faced by the grid operators, and also for investigation of causes of the blackouts. Next chapter is devoted to derivation of ANT.

References

1. B.B. Mandelbrot, *Fractals: Form, Chance, and Dimension*, Reprint ed. (Echo Point Books & Media, February 2020)
2. List of fractals by Hausdorff dimension. Available online. Accessed Feb. 2023 https://en.wikipedia.org/wiki/List_of_fractals_by_Hausdorff_dimension
3. J.L. Vevel, E. Lutton, C. Tricot (eds.), *Fractals in Engineering, From Theory to Practical Applications* (Springer, London, 1997)

Chapter 2
Active Network Twins

2.1 Introduction

Active Network Twins (ANT) is a mathematical phrase made up of parameters and topology of the power circuit. It is used as a linear multiplier in computation of line-flows and voltages in the power grid, and emulates the DNA in human body that determines shape and behaviours of various organs. This chapter discusses similarities between ANT and DNA. Consider a power network with *one* generator, with loads represented by impedances, i.e. $Z_L = 1/(P_L - jQ_L)$. Construct impedance parameter matrix *Z-net* for the network with ground-bus included. *Z-net* is different from the usual *Z-bus* because of inclusion of loads as impedances. Nodal admittances can also be used to obtain *Z-net*. With node-element matrix denoted by A,

$$[Y-\text{net}] = [A][y][A]^T$$

y is primitive admittance matrix including load admittances.

The last row and column correspond to ground node. The row and column have one load admittance each as off-diagonal entry with their sum as the diagonal corresponding to the ground node. *Y-net* is singular. Its Schur complement, obtained by eliminating ground node is a non-singular matrix which can be inverted to get *Z-net* (one dimension less than *Y-net*) with elements Z_{ki} corresponding to injection at bus i. We will denote *Z-net* by Z in the equations hereafter. For an element e with terminal nodes m and n, let

$$\xi_{ei} = [Z_{mi} - Z_{ni}] \tag{2.1}$$

Voltage across element e, with only ith generator connected to the circuit is,

$$v_{ei} = [A^T Z]_{\substack{\text{row},e \\ \text{col},i}} I_{ii},$$

© The Author(s), under exclusive license to Springer Nature Singapore Pte Ltd. 2023 7
S. D. Varwandkar and M. V. Hariharan, *Fractal Tomography for Power Grids*,
SpringerBriefs in Computational Intelligence,
https://doi.org/10.1007/978-981-99-3443-0_2

or,

$$v_{ei} = \xi_{ei} I_{ii} \tag{2.2}$$

I_{ii} is the generator current for the one-generator system. Equation (2.1) can be written for all circuit elements compactly as $\xi_{ei} = [A^T Z]_{\text{row},e}^{\text{col},i}$ where e is index of the circuit element between nodes m and n, and A is the node-element incidence matrix with ground node included. With $i_{ei} = y_e v_{ei}$, complex power in the element e is,

$$p_{ei} + jq_{ei} = v_{ei} i_{ei}^* \tag{2.3}$$

A little manipulation leads to,

$$p_{ei} + jq_{ei} = (\xi_{ei} \xi_{ei}^* y_e^*) I_{ii}^2 \tag{2.4}$$

2.2 ANT and DNA

The term $\xi_{ei} \xi_{ei} y_e^*$ in (2.4) represents a structure like the DNA (Fig. 2.1). This is a mathematical phrase, and is defined as *Active Network Twins*.

Note: DNA is a double-helix structure with two strands of sugar phosphates. Bases of the strands are nucleotides adenine, thymine, guanine and cytosine which bridge the two strands. Layers of genes are situated in layers over certain stretches of the DNA structure. Figure 2.1 is an attempt to depict Eq. (2.4). For power systems, I_{ii}^2 serves as gene, and the network parameters as nucleotides. ANT can be viewed as DNA with ξ_{ei} and ξ_{ei}^* as its strands.

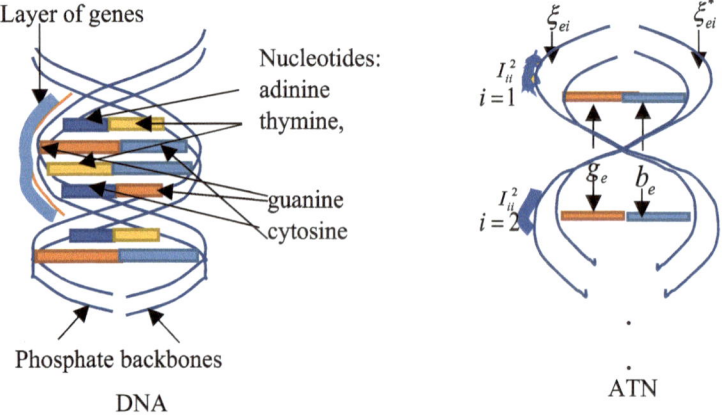

Fig. 2.1 DNA and ANT

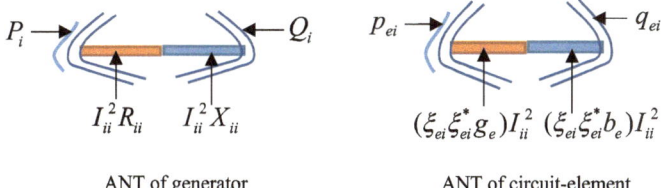

ANT of generator   ANT of circuit-element

Fig. 2.2 Variant of active network twins

Another variant of ANT with *power* as gene is shown in Fig. 2.2. This is more convenient to use when P and Q are known.

Let $Z_{ii} = R_{ii} + jX_{ii}$ be the diagonal element of Z (*Z-net*). It denotes driving point impedance. From power balance,

$$P_i + jQ_i = I_{ii}^2(R_{ii} + jX_{ii}) \tag{2.5}$$

I_{ii}^2 can be obtained from the power of the generator which is assumed constant.

$$I_{ii}^2 = \frac{P_i}{R_{ii}} \tag{2.6}$$

ANT and I_{ii}^2 (or Pi) determine electrical properties of all circuit elements of the system. Generator powers in various elements are isolated from network structure that determines the ANT with ξ_{ei} of (2.4) embedded into it. Equation (2.4) is of great advantage in separating influence of generators from that of the network in overall system performance. The network-based multiplier ANT can be tabulated and stored offline.

2.3 Square Domain

Sinusoidal shape of current waveform and its square are shown in Fig. 2.3 (scale exaggerated).

I-squares are always positive pulses unlike the continuously varying *alternating current*. I^2 converts the Kirchhoff's linear domain into a nonlinear *square domain*. Power transfer can be analysed using linear or a nonlinear domain. Nonlinearities of KCL variables become linearities in the square-domain. Variables v^2 and i^2 are square-domain voltage (*SD-voltage*) and square-domain current (*SD-current*), respectively. These have only magnitudes, no angles. Tomographs use SD-variables.

Fig. 2.3 Squaring the wave (scale adjusted)

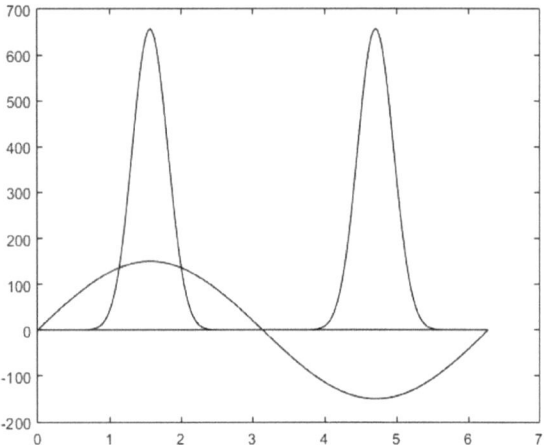

2.3.1 Example

Consider example shown in Fig. 2.4. Discrete power pulses of multiple generators in circuit elements can be obtained using (2.6), as follows.

Input of 100 MW remaining constant, power balance implies that pulses in 40/3 Ω must be 3 times those in the 40-Ω resistance 25 MW in 40 Ω and 75 MW in 40/3-Ω. Same ratio holds for the 200 MW source, i.e. 75 MW in 40 Ω and 225 MW in 40/3 Ω. To obtain load powers, power contributions in each resistance are summed. It may be noted that power balance principle is applied, not the KCL! In computing powers, we have used SD-variables. It simplifies calculations significantly. Detailed calculations are shown below.

1. $Z = Z-\text{net} = [(40/3)||40] = [10]$

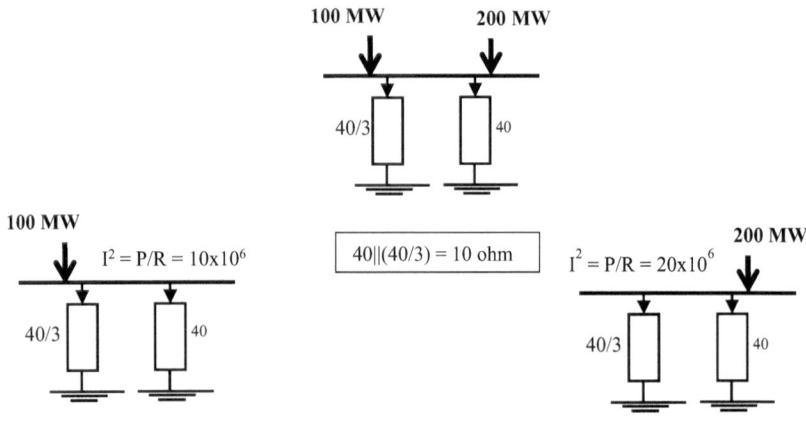

Fig. 2.4 *I-square* pulses of generators

2. $I_{11}^2 = \frac{100*10^6}{10} = 10*10^6$; $I_{22}^2 = \frac{200*10^6}{10} = 20*10^6$

    from to (ground)

3. $A^T = $ element 1 $\begin{bmatrix} 1 & -1 \\ 1 & -1 \end{bmatrix}$
    element 2

By ignoring second column of Z since the ground has zero potential,

4. $\xi_{ei} = [A^T Z]_{ei}$; $\xi_{11} = [A^T Z]_{11} = 10$; $\xi_{21} = [A^T Z]_{21} = 10$

5. $(ANT)_{11} = \xi_{11}\xi_{11}^* y_1^* = 10*10*\frac{3}{40} = 7.5$; $(ANT)_{21} = \xi_{21}\xi_{21}^* y_2^* = 10*10*\frac{1}{40} = 2.5$

6. $(ANT)_{12} = \xi_{12}\xi_{12}^* y_1^* = 10*10*\frac{3}{40} = 7.5$; $(ANT)_{22} = \xi_{22}\xi_{22}^* y_2^* = 10*10*\frac{1}{40} = 2.5$

    Power in $40/3\ \Omega = (ANT)_{40/3,1} * I_{11}^2 + (ANT)_{40/3,2} * I_{22}^2$

7. $\qquad\qquad = 7.5*10 + 7.5*20$

    $\qquad\qquad = 225\,MW$

    Power in $40\ \Omega = (ANT)_{40,1} * I_{11}^2 + (ANT)_{40,2} * I_{22}^2$

8. $\qquad\qquad = 2.5*10 + 2.5*20$.

    $\qquad\qquad = 75\,MW$

ANT does not change with generator injections. Following changes take place in analysis.

Domain	Linear domain	Square domain
Invariants	Specified bus-powers and specified generator-voltages	Generator (turbine)-powers
Model variables	Bus-voltage and angles	SD-voltage and currents
Analysis	KCL variables (equations)	Direct power equations
Model	Jacobian model (partial derivatives required)	Descriptor model (scalar) no partial derivatives
Model dimensions	Number of buses × Number of buses	Number of generators × Number of number of elements (lines + loads)
System matrix	Square, full rank	Rectangular with unitary property
Null space	Space of dependent variables	Independent and dependent variables, both are included in the descriptor model
Distribution factors	Requires voltage-angles of the base-case power flow	Scalar coefficients constitute fractals and can be calculated offline; base case is not required
Dynamics	Angles *evolve in time* starting from initial conditions	Different generator-frequencies *chase a final frequency* of rendezvous
Optimization	Physical meaning of objective function may not always be clear	Maximum power transfer is easily interpreted

(continued)

(continued)

Domain	Linear domain	Square domain
Shortcoming/ advantage	Analysis of cascading failures due to violently changing conditions cannot be easily performed	Cascading failures are easy to analyze
Convergence	Always an issue	Not relevant

2.4   Fractal Tomographs

Fractal tomograph (FT) is an ensemble of numerics calculated from parameters and topology of a power network and the turbine powers. It captures variables and functions in power systems like a CT scan does for a human body, spots any abnormality in lineflows and voltages, and can pinpoint its cause. Excessive lineflows and load pockets (low voltage areas) can be identified. Corrective actions are hinted in the process itself of computing the tomograph.

2.4.1   Structure of FT

Consider Fig. 2.5 the outer circles denote generators with their turbine powers, P_1, P_2, . . . , etc. The circle at the centre denotes a circuit element $r_e + jx_e$ or $g_e + jb_e$ which can be a transmission line or a load. Smaller double circles with multiple radial arrows between the two convey that one out of many parts of the generator power is used up by the circuit element. There are *as many* such arrows for an element as number of generators. Total used-up power in an element is sum of fractal powers of ng generators. Collection of rows of these fractals is called tomograph of the grid. Large tables representing fractal tomographs will be found in many places in this book.

2.4.2   Inversion

Let ng and ne be number of generators and of circuit elements, respectively. Tomograph in Fig. 2.5 is *invertible* in that, each *one* of the outer generator circles can also be treated as an inner circle, considered one at a time, surrounded by outer circles of ne circuit elements. Number of incoming arrows for each generator in the latter case is ne. Generator power is sum of ne element powers. The inverted picture represents tomograph for a generator. This is akin to physically inverting a collapsible hollow with ng-polygon frame at one end and ne-polygon at the other, inside out.

Fig. 2.5 Tomograph of an element

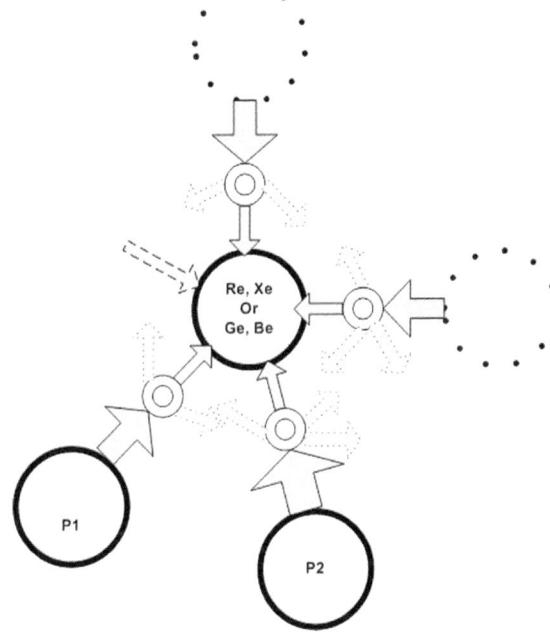

2.4.3 Mathematical Basis

In power systems, vector of turbine powers is ng and that of element-powers is ne. The $ne \times ng$ connection matrix for these two differently dimensioned vectors is non-square but satisfies property of a *unitary* matrix in that its transpose acts like its inverse. This can be verified by observation. We can consider Fig. 2.5 as several *one*-generator systems (Fig. 2.6).

Power balance equation for the generator can now be written as,

$$P_i = I_{ii}^2 R_{ii} \qquad (2.7)$$

The term R_{ii} is real part of the diagonal element $R_{ii} + jX_{ii}$ of Z-parameter matrix (the driving point impedance). From this, element powers are derived using traditional circuit laws. The resulting expression is multiplication of two numbers; one is a constant function of circuit parameters and the other, current-square of the

Fig. 2.6 Tomograph of a generator

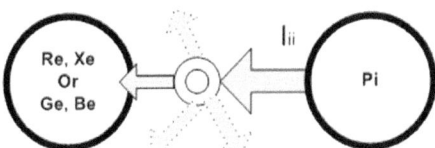

generator. Fractals of a generators in an element can thus be obtained. The sum of fractals gives net power in the element from which voltages across load elements can be determined. The ensemble of all fractals is *fractal tomograph*.

Chapter 3
Fractals and Tomographs

3.1 Introduction

Fractals represent patterns which can be visualized with an artist's eye. MWs and voltages of circuit elements are numbers which have no physical shapes but, like an artist, one can assign shapes to these variables and scale them as well. The shapes may not be useful for numeric manipulation; nevertheless, they lend themselves to comparison with natural fractals, especially when several of them come together to form a new one. Theory of fractals by Mandelbrot was inspired by natural repetitive shapes, but the theoretical development employed functions of a complex number which makes the interpretation more difficult. Natural reality is lost. Power system performance consists of real flows and real voltage magnitudes possess visibility and are meaningful to grid operators.

3.2 Lineflow Fractals

The node-element matrix (including ground bus) for a 4-bus 5-element network (figure not shown) with transmission line elements, 1–3, 2–3, 1–4, 2–4 and 3–4, and load impedances at all buses (4 circuit elements) has a pattern,

$$
\begin{array}{c}
\text{Node index} \downarrow \\
\begin{array}{c}
1 \\
2 \\
A = \quad 3 \\
4 \\
\text{ground}
\end{array}
\begin{array}{cccccccccc}
1 & 2 & 3 & 4 & 5 & 6 & 7 & 8 & 9 \\
\left[\begin{array}{ccccccccc}
1 & . & 1 & . & . & 1 & . & . & . \\
. & 1 & . & 1 & . & . & 1 & . & . \\
-1 & -1 & . & . & 1 & . & . & 1 & . \\
. & . & -1 & -1 & -1 & . & . & . & 1 \\
. & . & . & . & . & -1 & -1 & -1 & -1
\end{array}\right]
\end{array}
\end{array}
\tag{3.1}
$$

S. D. Varwandkar and M. V. Hariharan, *Fractal Tomography for Power Grids*, SpringerBriefs in Computational Intelligence, https://doi.org/10.1007/978-981-99-3443-0_3

A can be written as $A_1 - A_2$ with a single entry of '1' in each column of A_1 and A_2. For the given network,

$$
A_1 = \quad
\begin{array}{c}
\text{Node index} \\
\\
1 \\
2 \\
3 \\
4 \\
\text{ground}
\end{array}
\quad
\begin{array}{c}
\text{Element index} \\
\downarrow \quad 1 \ 2 \ 3 \ 4 \ 5 \ 6 \ 7 \ 8 \ 9 \\
\left[\begin{array}{ccccccccc}
1 & . & 1 & . & . & 1 & . & . & . \\
. & 1 & . & 1 & . & . & 1 & . & . \\
. & . & . & . & 1 & . & . & 1 & . \\
. & . & . & . & . & . & . & . & 1 \\
. & . & . & . & . & . & . & . & .
\end{array}\right]
\end{array}
\tag{3.2}
$$

$$
A_2 = \quad
\begin{array}{c}
\text{Node index} \\
\\
1 \\
2 \\
3 \\
4 \\
\text{ground}
\end{array}
\quad
\begin{array}{c}
\text{Element index} \\
\downarrow \quad 1 \ 2 \ 3 \ 4 \ 5 \ 6 \ 7 \ 8 \ 9 \\
\left[\begin{array}{ccccccccc}
. & . & . & . & . & . & . & . & . \\
. & . & . & . & . & . & . & . & . \\
1 & 1 & . & . & . & . & . & . & . \\
. & . & 1 & 1 & 1 & . & . & . & . \\
. & . & . & . & . & 1 & 1 & 1 & 1
\end{array}\right]
\end{array}
\tag{3.3}
$$

The term $[Z_{mi} - Z_{ni}]$ is "row e, column i" element of the $A^T Z$ matrix, e being the index of line element e *from*-bus m *to*-bus n ($e=3$ means $m=1$, $n=4$). A is the node-element incidence matrix. Used-power equation for lines can be written by splitting (2.2) of Chap. 2 as,

$$
p_{ei} = \text{Re}\left(Z_{mi}[Z_{mi}^* - Z_{ni}^*]y_e^* - Z_{ni}[Z_{mi}^* - Z_{ni}^*]y_e^*\right)I_{ii}^2
\tag{3.4}
$$

First term in (3.4) is incoming power in line e, and the second is outgoing power. Incoming power is taken as the lineflow.

$$
p_{eif} = \text{Re}\left(Z_{mi}[Z_{mi}^* - Z_{ni}^*]y_e^*\right)I_{ii}^2
\tag{3.5}
$$

Or,

$$
p_{eif} = \text{Re}\left(Z_{mi}\xi_{ei}^* y_e^*\right)I_{ii}^2
$$

Or,

$$
p_{eif} = \text{Re}\left([A_1 Z]_{ei}\xi_{ei}^* y_e^*\right)I_{ii}^2
\tag{3.6}
$$

p_{eif} is lineflow fractal, A_1 is first part of the decomposed node-element matrix $A = A_1 - A_2$. Net scalar lineflows are,

$$
p_{ef} = \sum_{i=1}^{ng} p_{eif}
\tag{3.7}
$$

The summation is valid for *used-up* powers in the elements, too, i.e.,

$$p_e = \sum_{i=1}^{ng} p_{ei} \tag{3.8}$$

Equation (3.4) can be written in terms of injected powers as,

$$p_{ef} = \sum_{i=1}^{ng} \text{Re} \frac{\left(\xi 1_{ei} \xi_{ei}^* y_e^*\right)}{R_{ii}} P_i \tag{3.9}$$

Lineflow fractals are

$$p_{eif} = \left\{ \text{Re} \frac{\left(\xi 1_{ei} \xi_{ei}^* y_e^*\right)}{R_{ii}} P_i \right\}; \; \xi 1_{ei} = [A_1^T Z]_{ei} \tag{3.10}$$

3.3 Flowgates and Corridors

Flowgates and transmission corridors are special cutsets of transmission lines. Let one of these be C. Then flowgate or corridor power can be obtained from,

$$p_{Cf} = \sum_{e \in C} \sum_{i=1}^{ng} p_{eif} \tag{3.11}$$

This expression is especially useful for determining area-to-area power flows in large systems. Equation (3.5) is the flow tomograph for element e of the grid and is depicted in Fig. 3.1.

3.4 Voltage Fractals

Voltage across a load element is given by its impedance triangle, i.e.

$$v_e^2 = i_e^2 r_e^2 + i_e^2 x_e^2 \tag{3.12}$$

$$v_e^2 = p_e r_e + q_e x_e \tag{3.13}$$

$$v_e^2 = p_e r_e \left(1 + \frac{x_e^2}{r_e^2}\right) \tag{3.14}$$

Fig. 3.1 Lineflow
tomograph

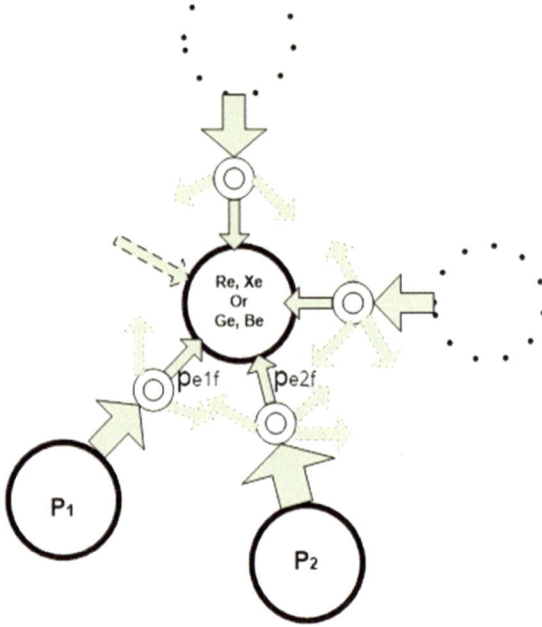

From (2.14) and (2.8), voltage fractals are obtained as,

$$\text{Voltage Fractals} = \left\{ r_e \left(1 + \frac{x_e^2}{r_e^2} \right) \frac{\xi_{ei} \xi_{ei}^* y_e^*}{R_{ii}} \right\} P_i \qquad (3.15)$$

Corresponding ANT is,

$$\text{ANT for voltage fractals} = \left\{ r_e \left(1 + \frac{x_e^2}{r_e^2} \right) \frac{\xi_{ei} \xi_{ei}^* y_e^*}{R_{ii}} \right\} \qquad (3.16)$$

3.5   Load Fractals

Load fractal is same as the power flow *into* the load element. The load tomograph is
therefore same as the *lineflow tomograph for load element.*

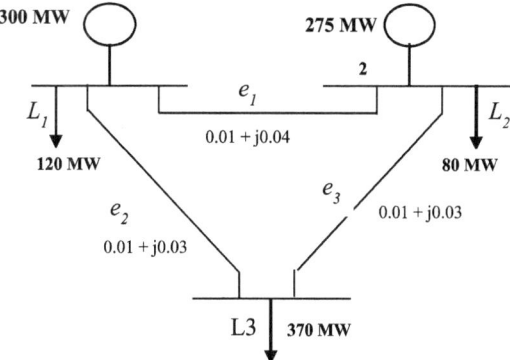

Fig. 3.2 Three-bus system

3.5.1 *Example*

Load fractals

Consider a three-bus power system as shown in Fig. 3.2.
The load fractals for the three loads are given by,

$$
\text{Load Fractals} = \begin{bmatrix} 0.2192 & 0.2063 \\ 0.1377 & 0.1468 \\ 0.6230 & 0.6241 \end{bmatrix} \begin{bmatrix} 300 \\ & 275 \end{bmatrix}
$$

$$
= \begin{bmatrix} 65.76 & 56.75 \\ 41.31 & 40.39 \\ 186.92 & 171.64 \end{bmatrix} \tag{3.17}
$$

Row sums give net load powers, i.e.,

$$
\text{Load powers} = \begin{bmatrix} 122.52 \\ 81.71 \\ 358.57 \end{bmatrix} \tag{3.18}
$$

Note that computed load powers are different than their specified values. This happens in order to maintain invariance of injected powers.

The fractals of used-up power (loss) in lines are given by (3.5) for line elements and found to be,

$$\text{Used up powers} = \begin{bmatrix} 1.096\,e^{-7} & 9.45\,e^{-8} \\ 1.03\,e^{-7} & 1.009\,e^{-7} \\ 1.01\,e^{-7} & 9.27\,e^{-08} \end{bmatrix} \begin{bmatrix} 300 \\ 275 \end{bmatrix}$$

$$= \begin{bmatrix} 2.04\,e^{-5} \\ 2.04\,e^{-5} \\ 1.94\,e^{-5} \end{bmatrix} \tag{3.19}$$

A pattern can be tagged to each fractal of (3.15) and (3.16) to obtain an artist's view depicted in Figs. 3.3 and 3.4. Fractals with colours make it look like a flowering plant. *Note: Repetitive occurrences in nature like this inspired the theory of fractals proposed by Mandelbrot but his theory uses complex numbers which deprives one of the feel of real objects.*

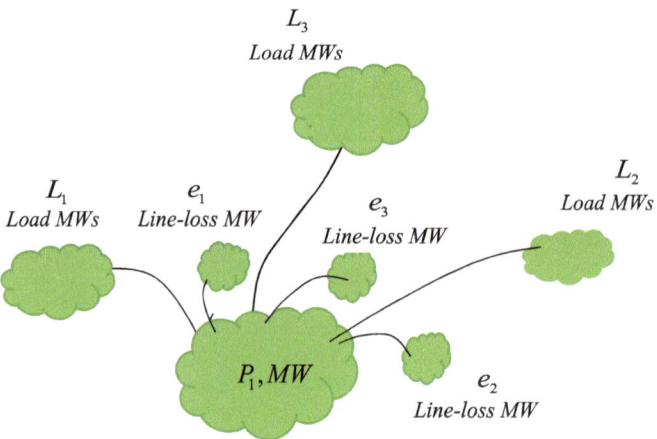

Fig. 3.3 Artist's view of load and line-loss fractals

Fig. 3.4 Tomograph for lineflow 1–3

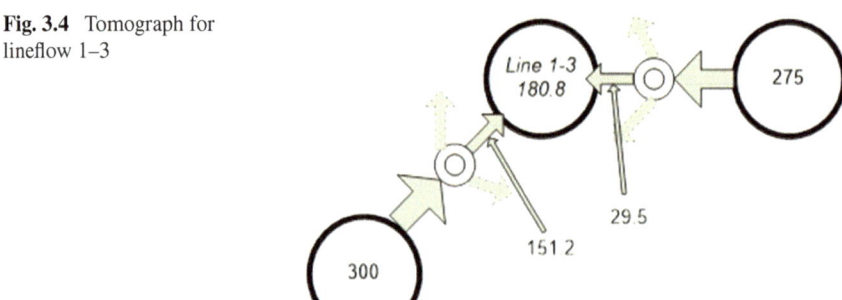

Voltage fractals

Voltage fractals can be obtained from (3.15).

$$\text{Voltage-fractals} = \begin{bmatrix} 0.5481 & 0.4729 \\ 0.5165 & 0.5049 \\ 0.5052 & 0.4639 \end{bmatrix} \tag{3.20}$$

$$\begin{bmatrix} v_{L_1}^2 \\ v_{L_2}^2 \\ v_{L_3}^2 \end{bmatrix} = \text{Row-sums} = \begin{bmatrix} 1.021 \\ 1.0214 \\ 0.969 \end{bmatrix} \tag{3.21}$$

Lineflow fractals

Line-power displacement fractal matrix for this system is obtained as.

$$\text{Lineflow fractals} = \begin{bmatrix} 83.01 & -86.34 \\ 151.21 & 29.59 \\ 40.35 & 146.35 \end{bmatrix} \tag{3.22}$$

Net power flows on transmission lines are given by row sums. For lines 1–2, 1–3 and 2–3, net lineflows are row sums of (3.22), that is,

$$\text{Net MW flows on lines} = \begin{matrix} \text{Line } 1-2 \\ \text{Line } 1-3 \\ \text{Line } 2-3 \end{matrix} \begin{bmatrix} -3.33441 \\ 180.8146 \\ 186.9869 \end{bmatrix} \tag{3.23}$$

Tomograph for lineflow 1–3 is shown in Fig. 3.5.

Lineflows, line loss and load voltages can be shown by enhancing the artist's viewpoint of Fig. 3.4, to that shown in Fig. 3.5.

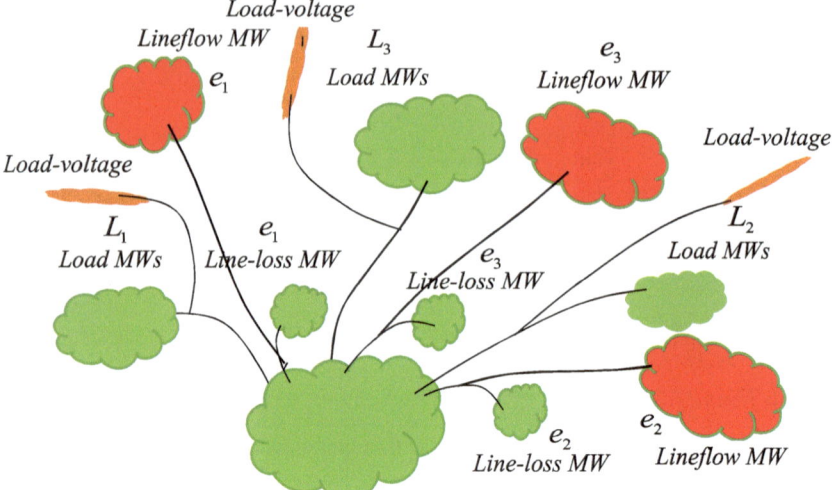

Fig. 3.5 Artist's view of loads, load voltages, lineflows and line-loss fractals

Chapter 4
Editing

4.1 Introduction

In frontier research in biological sciences, genes are edited to achieve desired performance in humans. Likewise, the fractal tomographs can be employed by system operators to control lineflows and load voltages during operation. Tomographs help them to edit appropriate fractals for intended changes in lineflows and voltages. Generator powers at specific buses are adjusted in editing. For every edit, a compensating power needs to be added or subtracted at another generator bus to preserve power balance. Desired changes can be identified all over the grid with only a few elementary calculations needed for achieving the objective. This chapter demonstrates this application on IEEE 39-bus New England system.

4.2 39-Bus New England System

Single line diagram of the 39-bus New England system is shown in Appendix along with data. The transmission line *form* 29-*to*-38 was renumbered as *from* 38-*to*-29 so that flow (maximum) which occurs on this line assumes *positive* value (original was *negative*, i.e. in opposite direction). Seventy percent of standard generator and load values are chosen as the base case. Base case load flow results are given in Table 4.1. Line # 45 (Table 4.1, line 38–29) has maximum flow 568.63 MW (Table 4.1, line 38–29). Limit being 500 MW, line 38–29 is congested.

S. D. Varwandkar and M. V. Hariharan, *Fractal Tomography for Power Grids*,
SpringerBriefs in Computational Intelligence,
https://doi.org/10.1007/978-981-99-3443-0_4

Table 4.1 Lineflows

From	To	MW
1	2	−49.0298
1	39	48.96313
2	3	182.6728
2	25	−73.6594
3	4	−18.4417
3	18	2.357694
4	5	−141.432
4	14	−163.639
5	6	−338.407
5	8	193.2371
6	7	265.0121
6	11	−210.727
7	8	131.2527
8	9	28.45179
9	39	17.45181
10	11	220.5726
10	13	233.4638
13	14	229.1976
14	15	58.29182
15	16	−139.962
16	17	217.0125
16	19	−307.932
16	21	−221.779
16	24	−36.442
17	18	99.06261
17	27	108.4057
21	22	−405.765
22	23	40.4031
23	24	242.3885
25	26	92.63444
26	27	84.73094
26	28	−31.9854
26	29	−68.6531
28	29	−219.272
12	11	−2.2708
12	13	−1.2239
6	31	−394.282

(continued)

Table 4.1 (continued)

From	To	MW
10	32	−454.306
19	33	−408.434
20	34	−341.845
22	35	−454.575
23	36	−384.948
25	37	−367.646
2	30	−174.757
38	**29**	**568.6303***
19	20	108.7109

*Overloaded line

4.3 Editing

Table 4.2—lineflow fractals can be viewed as numeric equivalent of CT scan for the power network. An example for blood cancer (from the net) is shown in Fig. 4.1 which shows a lump, like the large number '574.97' (for line 38–29) in Table 4.2.

Just as medical images can detect malignancy, the lineflow tomograph can detect and identify the cause of overloading of lines from its fractals. Fractal of generator 9 is maximum for line 38–29 (Table 4.2). Generator 9 is thus immediately spotted as the cause of overloading. Decongestion can be brought about by editing this fractal. The fractal, i.e. 574.97, is product of P_9 and its ANT. Power P_9 therefore needs to be changed. Let the MW to be subtracted from P_9 be X MW which should be added, preferably at bus with minimum ANT, i.e. generator 8 so as to have minimum opposite change in this line (Table 4.2). ANT are shown in Table 4.3.

Flow fractal of generator 9 in line 38–29 is 574.97 MW (Table 4.1). This fractal is reduced by, say X, to $574.97 - X$ in order to bring 38–29 lineflow close to 500 MW, and the fractal for generator 8 (with minimum ANT, Table 4.3) increased to $-0.858 + X$. Generation data which is used is 70% of the standard values. Calculations are:

$$(*)\ 3.99 * P_1 + 4.0 * P_2 + 4.0 * P_3 + 4.0 * P_4 + 4.0 * P_5 + 4.0 * P_5 + 4.0$$
$$* P_6 + 4.0 * P_7 + 3.99 * (P8 + X) + 6.36 * (P_9 - X) + 4.0 * P_{10} = 500$$

That gives,

Table 4.2 Lineflow fractal tomograph

Line		G1	G2	G3	G4	G5	G6	G7	G8	G9	G10
1	2	−33.966	1.037	−7.793	−29.869	−20.638	−33.641	−28.009	−55.613	−50.164	209.628
1	39	33.963	−1.043	7.786	29.864	20.635	33.636	28.004	55.607	50.158	−209.646
2	3	88.594	−38.496	−48.729	−33.774	−23.337	−38.040	−31.671	140.137	64.011	103.978
2	25	51.310	39.502	40.881	3.427	2.368	3.859	3.213	−197.335	−115.352	94.467
3	4	38.006	−108.333	−117.791	14.234	9.835	16.031	13.347	64.920	60.815	−9.508
3	18	37.896	49.170	45.261	−66.896	−46.222	−75.344	−62.730	54.010	−18.027	85.239
4	5	13.893	−154.720	−84.100	27.430	18.953	30.894	25.722	24.936	31.249	−75.687
4	14	10.508	8.477	−76.522	−39.671	−27.411	−44.681	−37.201	15.715	1.373	25.775
5	6	5.640	−207.264	−140.654	−2.903	−2.006	−3.269	−2.722	9.315	7.244	−1.788
5	8	8.207	50.394	56.002	30.234	20.890	34.052	28.351	15.538	23.891	−74.322
6	7	7.520	87.163	78.875	28.834	19.923	32.475	27.038	14.288	22.010	−53.113
6	11	−2.096	102.801	−220.686	−32.163	−22.223	−36.225	−30.160	−5.358	−15.215	50.599
7	8	1.690	66.418	58.165	17.334	11.977	19.523	16.254	3.908	9.913	−73.929
8	9	−3.129	72.405	69.296	22.082	15.258	24.871	20.707	−3.700	6.897	−196.235
9	39	−3.149	71.172	68.248	21.915	15.142	24.683	20.550	−3.718	6.874	−204.265
10	11	2.000	−92.553	226.213	29.675	20.504	33.423	27.827	5.057	14.106	−45.679
10	13	−2.023	92.482	228.251	−29.721	−20.536	−33.474	−27.870	−5.098	−14.154	45.607
13	14	−2.308	100.522	226.027	−32.806	−22.667	−36.948	−30.763	−5.744	−15.707	49.591
14	15	8.172	107.993	144.833	−72.802	−50.303	−81.996	−68.268	9.931	−14.376	75.108
15	16	0.067	86.833	118.825	−95.373	−65.898	−107.417	−89.433	−5.359	−33.299	51.092
16	17	−42.763	−8.807	5.903	148.068	102.308	166.766	138.846	−87.658	−136.912	−68.739

(continued)

Table 4.2 (continued)

Line		G1	G2	G3	G4	G5	G6	G7	G8	G9	G10
1	2	−33.966	1.037	−7.793	−29.869	−20.638	−33.641	−28.009	−55.613	−50.164	209.628
16	19	13.253	29.344	34.544	−330.757	−228.538	43.511	36.226	25.460	32.013	37.012
16	21	9.526	21.092	24.829	27.768	19.186	−237.771	−154.322	18.300	23.010	26.603
16	24	11.574	25.627	30.168	33.739	23.312	−108.992	−134.386	22.235	27.958	32.324
17	18	−32.735	−39.197	−33.881	77.985	53.884	87.833	73.128	−44.862	28.746	−71.840
17	27	−10.392	30.367	39.767	68.378	47.246	77.013	64.119	−43.610	−167.232	2.751
21	22	2.639	5.843	6.879	7.693	5.315	−272.574	−180.375	5.070	6.375	7.370
22	23	2.625	5.812	6.841	7.651	5.287	176.761	−183.286	5.042	6.340	7.330
23	24	−3.561	−7.885	−9.283	−10.381	−7.173	141.148	164.913	−6.842	−8.602	−9.946
25	26	39.723	24.882	24.325	−9.611	−6.640	−10.824	−9.012	114.316	−145.410	70.885
26	27	18.754	−13.658	−20.277	−49.530	−34.223	−55.784	−46.445	62.792	204.225	18.876
26	28	8.562	16.488	19.113	17.153	11.852	19.320	16.085	20.418	−182.775	21.798
26	29	6.984	13.449	15.590	13.992	9.668	15.759	13.120	16.654	−191.649	17.780
28	29	2.010	3.871	4.487	4.027	2.782	4.535	3.776	4.793	−254.669	5.117
12	11	0.116	−8.965	−0.416	2.683	1.854	3.022	2.516	0.340	1.183	−4.605
12	13	−0.266	8.485	−0.226	−2.991	−2.066	−3.368	−2.804	−0.610	−1.500	4.123
6	31	0.201	−398.177	0.735	0.400	0.276	0.450	0.375	0.359	0.419	0.680
10	32	0.011	0.035	−454.517	0.023	0.016	0.026	0.021	0.020	0.023	0.036
19	33	0.738	1.633	1.923	−425.541	3.116	2.422	2.016	1.417	1.782	2.060
20	34	0.073	0.162	0.191	0.447	−343.680	0.240	0.200	0.141	0.177	0.204

(continued)

Table 4.2 (continued)

Line		G1	G2	G3	G4	G5	G6	G7	G8	G9	G10
1	2	−33.966	1.037	−7.793	−29.869	−20.638	−33.641	−28.009	−55.613	−50.164	209.628
22	35	0.006	0.013	0.016	0.017	0.012	−454.709	0.027	0.011	0.014	0.017
23	36	0.000	0.000	0.000	0.000	0.000	0.000	−384.948	0.000	0.000	0.000
25	37	0.139	0.227	0.261	0.219	0.152	0.247	0.206	−369.743	0.328	0.318
2	30	−174.867	0.012	0.014	0.011	0.007	0.012	0.010	0.013	0.013	0.018
38	**29**	−0.360	−0.693	−0.804	−0.721	−0.498	−0.812	−0.676	−0.858	**574.970***	−0.916
19	20	13.850	30.666	36.101	84.666	−238.645	45.471	37.859	26.608	33.456	38.680

* Largest fractal belongs to generator 9

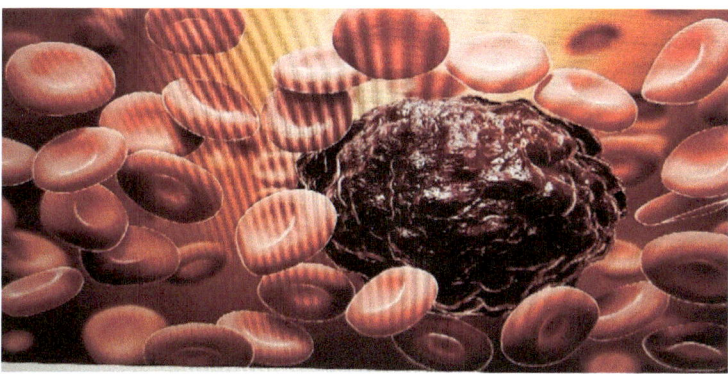

Fig. 4.1 Medical image (CT) of blood cancer

$$X = 68.5$$

With modified powers, lineflow on 38–29 comes down to 499.8435 MW. Other fractals are marginally affected. Edited lineflows are given in Table 4.4.

ANT are constants and make calculations easier. Only two generators may be sufficient for editing although more can also be considered. Implementation is easier for two. Surgical editing (load shedding) is required occasionally. But it will add one repeat calculation for each surgical action. *The purpose of showing full tables is to emphasize the fact that a small targeted change in the tomograph can achieve desired performance in the grid.* Voltages can also be edited in a similar manner. This is left as an exercise for readers who may like to refer [1] (Table 4.5) for guidance.

Table 4.3 Active network twins for lineflows ($\times 10^{-6}$)

Element		Generators									
From	To	1	2	3	4	5	6	7	8	9	10
1	2	3.50	4.01	3.96	3.86	3.88	3.84	3.84	3.64	3.79	4.91
1	39	4.50	3.99	4.04	4.14	4.12	4.16	4.16	4.36	4.21	3.09
2	3	5.31	3.79	3.76	3.84	3.87	3.82	3.82	4.90	4.26	4.45
2	25	4.76	4.22	4.20	4.02	4.01	4.02	4.02	2.73	3.53	4.41
3	4	4.56	3.40	3.41	4.07	4.05	4.08	4.08	4.42	4.25	3.96
3	18	4.56	4.27	4.23	3.69	3.74	3.64	3.64	4.35	3.93	4.37
4	5	4.21	3.15	3.58	4.13	4.11	4.15	4.15	4.16	4.13	3.67
4	14	4.16	4.05	3.62	3.82	3.85	3.79	3.79	4.10	4.01	4.11
5	6	4.08	2.86	3.30	3.99	3.99	3.98	3.98	4.06	4.03	3.99
5	8	4.12	4.28	4.28	4.14	4.12	4.16	4.16	4.10	4.10	3.68
6	7	4.11	4.48	4.39	4.13	4.11	4.16	4.15	4.09	4.09	3.77
6	11	3.97	4.57	2.90	3.85	3.88	3.83	3.83	3.97	3.94	4.22
7	8	4.02	4.37	4.29	4.08	4.07	4.09	4.09	4.03	4.04	3.68
8	9	3.95	4.40	4.34	4.10	4.09	4.12	4.12	3.98	4.03	3.15
9	39	3.95	4.39	4.34	4.10	4.08	4.12	4.12	3.98	4.03	3.11
10	11	4.03	3.49	5.13	4.14	4.11	4.16	4.16	4.03	4.06	3.80
10	13	3.97	4.51	5.14	3.86	3.89	3.84	3.84	3.97	3.94	4.20
13	14	3.97	4.55	5.12	3.85	3.87	3.82	3.82	3.9 6	3.94	4.22
14	15	4.12	4.59	4.72	3.66	3.72	3.61	3.61	4.06	3.94	4.33
15	16	4.00	4.48	4.59	3.56	3.63	3.48	3.49	3.97	3.86	4.22
16	17	3.37	3.95	4.03	4.69	4.57	4.80	4.79	3.43	3.44	3.70
16	19	4.20	4.16	4.17	2.46	2.72	4.21	4.21	4.16	4.13	4.16
16	21	4.14	4.12	4.12	4.13	4.11	2.86	3.12	4.12	4.09	4.12
16	24	4.17	4.14	4.15	4.16	4.13	3.48	3.23	4.14	4.11	4.14
17	18	3.52	3.78	3.83	4.36	4.30	4.42	4.42	3.71	4.12	3.69
17	27	3.85	4.17	4.20	4.32	4.26	4.37	4.37	3.72	3.31	4.01
21	22	4.04	4.03	4.03	4.04	4.03	2.69	2.97	4.03	4.03	4.03
22	23	4.04	4.03	4.03	4.04	4.03	4.85	2.96	4.03	4.03	4.03
23	24	3.95	3.96	3.95	3.95	3.96	4.68	4.94	3.96	3.96	3.96
25	26	4.59	4.14	4.12	3.96	3.96	3.95	3.95	4.74	3.40	4.31
26	27	4.28	3.92	3.90	3.77	3.81	3.73	3.74	4.41	4.84	4.08
26	28	4.13	4.09	4.10	4.08	4.07	4.09	4.09	4.13	3.25	4.09
26	29	4.10	4.07	4.08	4.07	4.05	4.08	4.07	4.11	3.21	4.08
28	29	4.03	4.02	4.02	4.02	4.02	4.02	4.02	4.03	2.95	4.02
12	11	4.00	3.95	4.00	4.01	4.01	4.01	4.01	4.00	4.00	3.98

(continued)

Table 4.3 (continued)

Element		Generators									
From	To	1	2	3	4	5	6	7	8	9	10
12	13	4.00	4.05	4.00	3.99	3.99	3.98	3.98	4.00	3.99	4.02
6	31	4.00	1.81	4.00	4.00	4.00	4.00	4.00	4.00	4.00	4.00
10	32	4.00	4.00	1.74	4.00	4.00	4.00	4.00	4.00	4.00	4.00
19	33	4.01	4.01	4.01	2.02	4.02	4.01	4.01	4.01	4.01	4.01
20	34	4.00	4.00	4.00	4.00	2.08	4.00	4.00	4.00	4.00	4.00
22	35	4.00	4.00	4.00	4.00	4.00	1.81	4.00	4.00	4.00	4.00
23	36	4.00	4.00	4.00	4.00	4.00	4.00	1.81	4.00	4.00	4.00
25	37	4.00	4.00	4.00	4.00	4.00	4.00	4.00	1.61	4.00	4.00
2	30	1.42	4.00	4.00	4.00	4.00	4.00	4.00	4.00	4.00	4.00
38	**29**	3.99	4.00	4.00	4.00	4.00	4.00	4.00	3.99	**6.36***	4.00
19	20	4.20	4.17	4.18	4.39	2.67	4.22	4.22	4.17	4.14	4.17

*ANT of largest fractal

Table 4.4 Edited lineflows

From	To	MW
1	2	−53.245
1	39	53.17803
2	3	200.7424
2	25	−96.0945
3	4	−13.7901
3	18	14.4183
4	5	−140.587
4	14	−160.919
5	6	−337.562
5	8	193.236
6	7	265.0063
6	11	−209.894
7	8	130.7864

(continued)

Table 4.4 (continued)

From	To	MW
8	9	26.94969
9	39	15.9492
10	11	219.8167
10	13	234.218
13	14	230.0187
14	15	61.82972
15	16	−136.971
16	17	217.2726
16	19	−307.082
16	21	−221.168
16	24	−35.6998
17	18	87.40095
17	27	120.3658
21	22	−405.596
22	23	40.57142
23	24	242.1602
25	26	130.9635
26	27	71.87431
26	28	−6.42319
26	29	−42.7221
28	29	−187.995
12	11	−2.3496
12	13	−1.15663
6	31	−394.266
10	32	−454.305
19	33	−408.387
20	34	−341.841
22	35	−454.574
23	36	−384.948
25	37	−435.519
2	30	−174.756
38	**29**	**499.8435***
19	20	109.599

*Edited MW on line 38–29 flow

Table 4.5 Generator powers

Generator	Power-injection, MW
1	175
2	401.05
3	455
4	442.4
5	355.6
6	455
7	392
8	378
9	**581***
10	700

* These values are 70% of the base case (Appendix)

4.4 Conclusion

Controlling lineflows and voltages is an important activity of grid operators to ensure security. Targeted editing can be used for lineflows. Fractal approach enables them to spot critical lines and edit its fractals. Obviously, all targets cannot be edited due to capability constraints or operational limits of power equipment. 'How many edits are possible?' is a question that requires further research.

APPENDIX

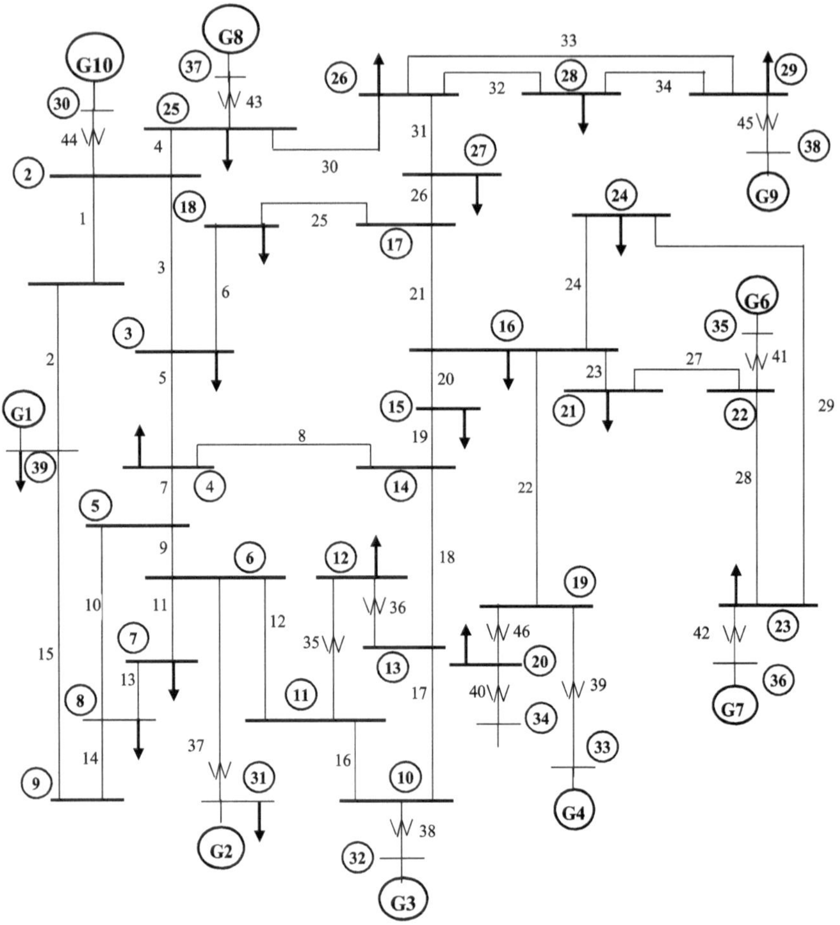

Fig. 4.2 39-bus New England system

Table 4.6 Line data

From	To	Line #	R	X	B	Tap-ratio
1	1	2	0.0035	0.0411	0.3493	1
2	1	39	0.001	0.025	0.375	1
3	2	3	0.0013	0.0151	0.1286	1
4	2	25	0.007	0.0086	0.073	1
5	3	4	0.0013	0.0213	0.1107	1
6	3	18	0.0011	0.0133	0.1069	1
7	4	5	0.0008	0.0128	0.0671	1
8	4	14	0.0008	0.0129	0.0691	1
9	5	6	0.0002	0.0026	0.0217	1
10	5	8	0.0008	0.0112	0.0738	1
11	6	7	0.0006	0.0092	0.0565	1
12	6	11	0.0007	0.0082	0.0694	1
13	7	8	0.0004	0.0046	0.039	1
14	8	9	0.0023	0.0363	0.1902	1
15	9	39	0.001	0.025	0.6	1
16	10	11	0.0004	0.0043	0.0364	1
17	10	13	0.0004	0.0043	0.0364	1
18	13	14	0.0009	0.0101	0.0862	1
19	14	15	0.0018	0.0217	0.183	1
20	15	16	0.0009	0.0094	0.0855	1
21	16	17	0.0007	0.0089	0.0671	1
22	16	19	0.0016	0.0195	0.152	1
23	16	21	0.0008	0.0135	0.1274	1
24	16	24	0.0003	0.0059	0.034	1
25	17	18	0.0007	0.0082	0.0654	1
26	17	27	0.0013	0.0173	0.1608	1
27	21	22	0.0008	0.014	0.1282	1
28	22	23	0.0006	0.0096	0.0923	1
29	23	24	0.0022	0.035	0.1805	1
30	25	26	0.0032	0.0323	0.2565	1
31	26	27	0.0014	0.0147	0.1198	1
32	26	28	0.0043	0.0474	0.3901	1
33	26	29	0.0057	0.0625	0.5195	1
34	28	29	0.0014	0.0151	0.1295	1
35	12	11	0.0016	0.0435	0.0001	1.006
36	12	13	0.0016	0.0435	0.0001	1.006
37	6	31	0	0.025	0.0001	1.11
38	10	32	0	0.02	0.0001	1.07

(continued)

Table 4.6 (continued)

From	To	Line #	R	X	B	Tap-ratio
39	19	33	0.0007	0.0142	0.0001	1.07
40	20	34	0.0009	0.018	0.0001	1.009
41	22	35	0	0.0143	0.0001	1.025
42	23	36	0.0005	0.0272	0.0001	1
43	25	37	0.0006	0.0232	0.0001	1.025
44	2	30	0	0.0181	0.0001	1.025
45	38	29	0.0008	0.0156	0.0001	0.9756
46	19	20	0.0007	0.0138	0.0001	1.06

Table 4.7 Base case load-generation data* (70% of standard values)

Loads				Generation		
Load No	At bus	Load MW	Load MVAR	Gen. No	At bus	Generation, MW
1	3	225.4	2.4	1	30	175
2	4	350	184	2	31	401.05
3	7	163.66	84	3	32	517.31
4	8	365.4	176	4	33	442.4
5	12	5.25	88	5	34	355.6
6	15	224	153	6	35	455
7	16	230.3	32.3	7	36	392
8	18	110.6	30	8	37	378
9	20	439.6	103	9	38	518.06
10	21	191.8	115	10	39	700
11	23	173.25	84.6			
12	24	216.02	−92			
13	25	156.8	47.2			
14	26	97.3	17			
15	27	196.7	75.5			
16	28	144.2	27.6			
17	29	198.45	26.9			
18	31	6.44	4.6			
19	39	772.8	250			

Reference

1. S. Varwandkar, J. Lin, Cost of voltage violation, in *9th International Conference on Power Systems (ICPS2021)*, 16–18 December 2021, IIT Kharagpur, India

Chapter 5
Power Reversals and Loop Flows

5.1 Introduction

Litigations sprung up between stakeholders and regulators in power markets in the first decade of this century as phenomenon of *Lake Erie Circulation* [1] in the US-Canada region played truant with contracted flows in power markets. This invited attention of system experts even as the trading had to be banned over certain routes. The term "loop flow" was coined around that time. It is used in a different sense than the "current loop" in electric circuits. Power flows that occur on paths other than the contracted ones were *defined* as loop flows. Actual flows on lines can get reversed because of schedules and can impact power trading [2]. Phase-angle regulators (PARs) were employed to avoid loop flows but were not quite effective. Their coordination was difficult. Consequently, PARs were operated mostly in the unregulated mode. Power *tracing* also became a topic of research. The proportionality principle used in *power tracing* was however analytically weak. Modifications were devised, but the method was essentially local. Tomography can determine power flows reliably and globally. It is analytically foolproof and can pinpoint generators responsible for causing loop flows and reversals. Vulnerable transmission lines can be identified. In this chapter, we analyse loop flows with tomographs. The method can be used to advantage by grid operators. A practical network is chosen for illustration.

5.2 Background of Loop Flows

Geography of US–Canada network region comprising PJM, MISO, IESO (Ontario) and NYISO is shown in Fig. 5.1. MISO (earlier METC and ITC) is connected to IESO (earlier IMO) via a loop formed by lines on the north and south of the Lake Erie. Substantial load in Ohio region (First Energy) is fed by the nuclear plants on the south-east of Lake Erie, and from MISO on its west. Market-cleared lineflows in a

particular case are shown in black in Fig. 5.2. Actual flows are shown in red. 1000 MW
are contracted by NYISO from AEP with transmission rights contracted on the route
AEP-MICHIGAN-IESO-NYISO as shown on Fig. 5.2. No trading was allowed on
the shorter route AEP-PJM-NYISO due to congested lines. The intermediate transmission contracts between adjacent nodes are shown by black arrows in Fig. 5.2.
Everything seems all right for the market. However, the power physically flows on
the low impedance path between AEP and NYISO thereby overloading the already
congested lines in this region. This is said to be due to the loop flows. A notional
network is shown in Fig. 5.3.

Flows shown in Fig. 5.2 can be marked on the transmission lines in Fig. 5.3.
Lineflow tomographs are used for this purpose.

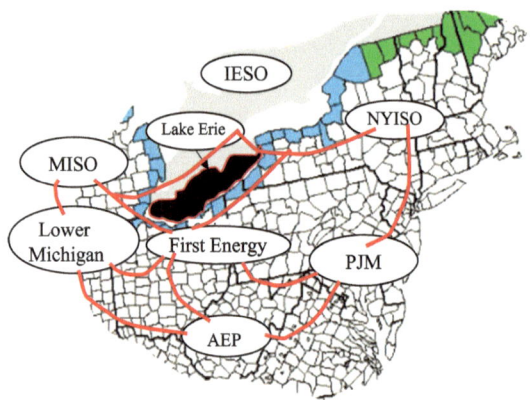

Fig. 5.1 US–Canada geography with Lake Erie circulation

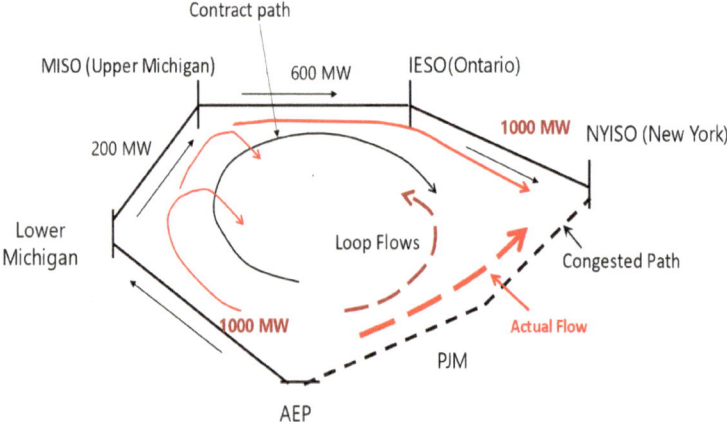

Fig. 5.2 Contracted and actual paths (buyer NYISO: seller AEP)

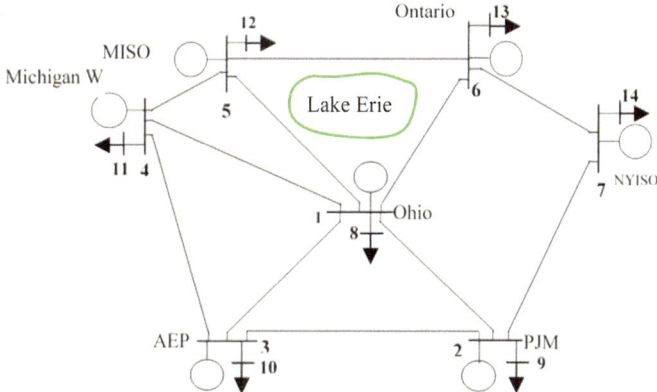

Fig. 5.3 Notional network

5.3 Lineflow Tomographs

Lineflow tomograph for the system can be written as,

$$
\begin{bmatrix}
p_{1f} \\
p_{2f} \\
p_{3f} \\
\cdot \\
\cdot \\
p_{ne,f}
\end{bmatrix}
=
\begin{bmatrix}
\varepsilon_{11f} P_1 & \varepsilon_{12f} P_2 & \cdots & \varepsilon_{1,ng,f} P_{ng} \\
\varepsilon_{21f} P_1 & \varepsilon_{22f} P_2 & \cdots & \varepsilon_{2,ng,f} P_{ng} \\
\cdot & \cdot & & \cdot \\
\cdot & \cdot & \cdots & \cdot \\
\cdot & \cdot & & \cdot \\
\varepsilon_{ne,1,f} P_1 & \varepsilon_{ne,2,f} P_2 & \cdots & \varepsilon_{ne,ng,f} P_{ng}
\end{bmatrix}
\tag{5.1}
$$

First subscript of ε refers to element e and the second, to the generator, 'f' is for flow. Expressions for lineflow fractals (3.10) are reproduced below.

$$
\mathrm{Re}\left(\frac{\xi 1_{ei} \xi_{ei}^* y_e^*}{R_{ii}} \right) P_i
\tag{5.2}
$$

Lineflow tomograph for the eth line is row e of (5.1) *wrapped-around* "p_{ef}" as shown in Fig. 5.4. It is interesting to compare Fig. 5.4 with Fig. 2.5 of Chap. 2 when drawn for flows.

Similarly, a *generator tomograph* is column i wrapped around the generator i as depicted in Fig. 5.5.

Fig. 5.4 Row e of Eq. 5.1 wrapped-around lineflow in *element e*

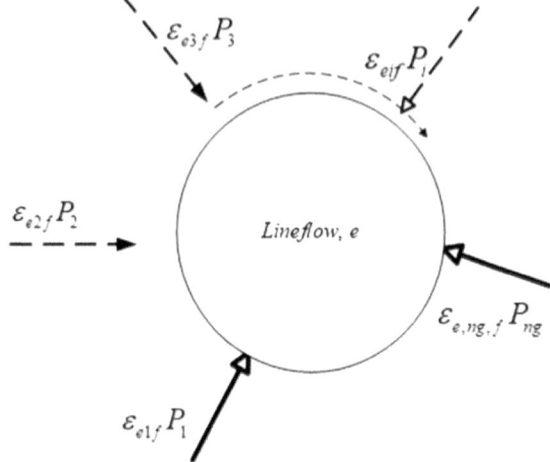

Fig. 5.5 Column i wrapped-around *generator i*

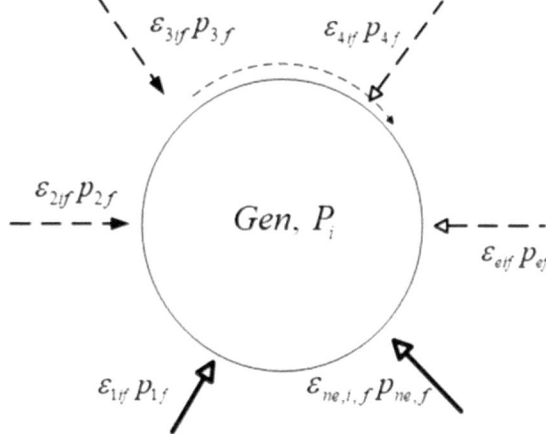

5.4   Example

Following are data tables for the notional network of Fig. 5.3 (Tables 5.1, 5.2, 5.3 and 5.4).

Lineflow tomographs are shown in Table 5.5. Generators are seen to contribute to lineflows in either direction signed as (+) or (-). Generator fractals with superscript 1 are positive and those with superscript 2, negative in Table 5.5. Contribution of (+) 1443.6 MW from OHIO generators in line 1–5 is *from* 1 *to* 5 but that of IESO is (−) 2059.9 MW (italicized in Table 5.5).

Table 5.6 summarizes net lineflows caused by new schedules (New case) indicating effect of the loop flows on lines 1–5, 4–5, 5–6 and 6–7. Base flow was − 1020.8 MW i.e. from 5 to 1 (Table 5.5), which reverses to 176.1 MW (Table 5.6). The

Table 5.1 Line data, per unit on 100 MVA base

From	To	R	X	B
1	2	0.0001	0.0004	0.00010
1	3	0.0001	0.0005	0.00010
1	4	0.0003	0.0010	0.00010
1	5	0.0001	0.0004	0.00010
1	6	0.0001	0.0004	0.00010
2	3	0.0001	0.0004	0.00010
2	7	0.0002	0.0008	0.00010
3	4	0.0001	0.0005	0.00010
4	5	0.0001	0.0004	0.00010
5	6	0.0001	0.0004	0.00010
6	7	0.0003	0.0010	0.00010
1	8 (1)	0.0003	0.0012	0.00001
1	8 (2)	0.0003	0.0012	0.00001
2	9 (1)	0.0001	0.0002	0.00001
2	9 (2)	0.0001	0.0002	0.00001
3	10 (1)	0.0001	0.0004	0.00001
3	10 (2)	0.0001	0.0004	0.00001
4	11 (1)	0.0001	0.0004	0.00001
4	11 (2)	0.0001	0.0004	0.00001
5	12 (1)	0.0001	0.0004	0.00001
5	12 (2)	0.0001	0.0004	0.00001
6	13 (1)	0.0001	0.0004	0.00001
6	13 (2)	0.0001	0.0004	0.00001
7	14 (1)	0.0001	0.0004	0.00001
7	14 (2)	0.0001	0.0004	0.00001

Table 5.2 Loads, MW

Bus	Ohio	PJM	AEP	MICH. west	MICH. east	IESO	NYISO
Load	13,635	63,263	28,341	20,660	8000	22,860	28,402

Table 5.3 Base-case generation, MW

Bus	Ohio	PJM	AEP	MICH. west	MICH. east	IESO	NYISO
Gen	9940	63,962	32,854	23,234	9600	26,844	32,015

Table 5.4 New schedule of generation, MW

Bus	Ohio	PJM	AEP	MICH. west	MICH. east	IESO	NYISO
Gen	10,940	63,650	27,246	17,225	8600	22,986	24,376

Table 5.5 Lineflow tomograph for selected lines, base case (in MW)

From bus →		1	4	5	6	Other lines
To bus →		5	5	6	7	
MW-Contributions from generators	Ohio	..	*1443.6*	-584.5^2	367.6	885	..
	PJM	..	4107.1	-340.2^2	1214.0	-544.8	..
	AEP	..	1044.3	2480.6^1	2136.5	1591.9	..
	MICH (W)	..	-2440.6	7529.2^1	3581.6	1896.7	..
	MICH (E)	..	-3128.5	-2688.6^2	2910.4	1134.1	..
	IESO	..	*-2059.9*	-3834.6^2	-7239.5	5224.2	..
	NYISO	..	13.152	-1716.6^2	-2780.1	-8875.7	..
Net flow, MW		..	-1020.8	$845.^{(1+2)}$	190.4	1311.9	.

Note Superscripts 1 and 2 denote generator groups with positive and negative displacements, respectively

difference of $+1196.9$ MW, as per definition, is the "loop flow". The Lake Erie circulation is due to strong *reversed* flow in line 1–5. Reversals are highlighted in Table 5.6. Table 5.7 shows generator fractals of on lines 6–7 and 2–7 when the schedules change. This may overload lines. The information is useful to system operator while revising schedules. Market operators can use it for amicably apportioning congestion charges among the customers. Figure 5.6 shows directions of original lineflows by normal arrows, new ones by bold arrows and loop flows by dotted arrows.

Table 5.6 Loop flows and reversals

From bus	1	4	5	6
To bus	5	5	6	7
Base-case flow, MW	-1020.8	845.1	190.5	1311.9
New case flow, MW	176.14	-341.9	331.2	1890.2
Loop flow, MW	1196.9	-1187.1	140.7	578.35
Reversal/addition of displacement	Strong reversal	Strong reversal	Addition	Addition

Table 5.7 Generator fractals of lineflows 6–7 and 2–7, MW

e/Gen	Ohio	PJM	AEP	MI-W	MI-E	OH	NYIS	Net
6–7	974	-542	1320	1406	1016	4473	-6757	1890
2–7	527	9795	2203	736	135	-945	-8978	3475

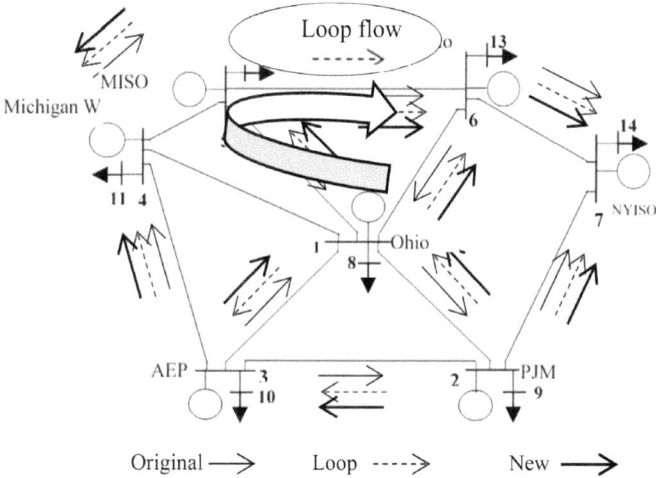

Fig. 5.6 Displacement loops and reversals

5.5 Phase-Angle Regulator

Phase-angle regulators (PARs) or phase shifting transformers can control lineflows and were suggested for mitigating loop flows. However, control on one line caused uncertain changes/reversals on some other lines. Coordination became difficult. Settings were therefore not permitted unless all PARs were involved in the process. Implementation of PARs has not been very successful in resolving the problem. Fractal tomographs can precisely compute effect of a PAR on all lines. Transformer is essentially a multiplier that changes the 'impedance seen by the transformer'. In ratio transformers, tap 't' adjusts *ratios* of *magnitudes*. In phase-angle regulators, *phase difference* 'α' is changed. PARs are usually rotary devices that do not change magnitudes but *add/subtract* an angle α *to/from* the bus-voltage-angle on the *other* side of the transformer. Physically, PAR changes the 'seen impedance from $r + jx$ to $r' + jx'$ such that, $r^2 + x^2 = r'^2 + x'^2$ but $\tan^{-1} \frac{x'}{r'} = \tan^{-1} \frac{x}{r} \pm \alpha$. PARs can be configured via transformer connections too (e.g. the Scott connection shifts phase by 90°). Impedances Z_{mi} and Z_{ni} are source-side *and* load-side impedances, respectively. PAR in line m–n adds an angle to the original angle of Z_{ni}, i.e.

$$\angle Z'_{ni} = (\angle Z_{ni} + \alpha) \tag{5.3}$$

Figure 5.7 gives geographical locations related to the event.

To determine α for desired change in power displacement, several values need to be tried, fractal displacements for each computed and the best value of α chosen. Direct computation is not possible due to intricate dependence of Z-bus parameters on the line/load data. To illustrate this, we introduce PAR with phase-shift $\alpha = \frac{0.5\pi}{180}$ rad to obtain Table 5.8.

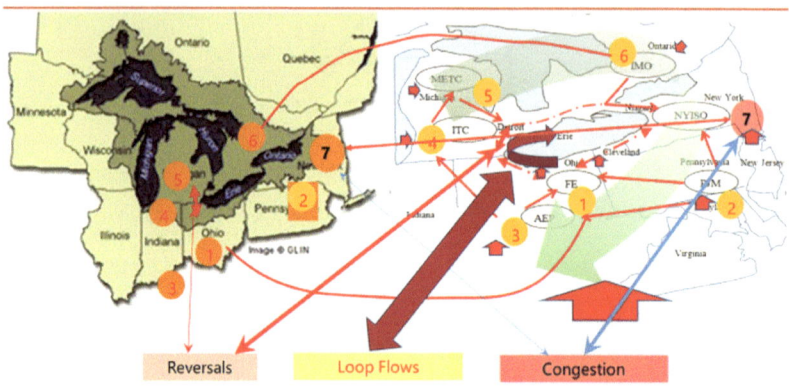

Fig. 5.7 Reversals, loop flows and congestions

Table 5.8 Selected lineflows with PAR

From bus	1	4	5	6
To bus	5	5	6	7
M W flow base case without PAR	−1020.8	845.1	190.5	1311.9
M W flow new case without PAR	**176.14**	**−341.9**	331.2	1890.2
M W flow new case with PAR (−0.5°)	**−1013.6**	**858.5**	2314.1	467.8
Reversal/relief/addition with respect to Base case	Almost unchanged	Almost unchanged	High Addition	High Flow-relief

It is seen that while a setting of (−) 0.5° can prevent (at least arrest) occurrence of reversal in line 1–5, it can however cause unusual changes on other lines (e.g. line 4–5).

5.6 Conclusion

Fractal tomography can be used to compute loop flows and analytically detect flow reversals with mathematical rigour. Phase-angle regulators (PARs) can be included in analysis. Cascading failures need intensive studies of power reversals and loop flows for which fractal tomography can be effectively used. Secure operation can be ensured and power grids made robust. Tomographs can amicably resolve conflicts in power markets. We shall illustrate this in greater details in Chap. 8.

References

1. Technical Report, Investigation of loop flows across combined Midwest ISO and PJM footprint, 25 May 2007
2. FERC Document 142 # 61202 [Available Online]

Chapter 6
Blackouts

6.1 Introduction

A tomographic study is presented in this chapter that provides a glimpse of intricacies of three important outage scenarios and provides valuable insights into shortcomings in the design of systems. Operational and planning deficiencies can be identified and remedied. The disturbances considered for illustration are: (i) energy-deficient distribution network in Mumbai, India, which imports 40–50% of its requirement from Maharashtra State Electricity Transmission Company Limited (MSETCL) and which experienced a blackout on October 12, 2020, (ii) an export-rich network of Bhutan which is precisely the opposite case with an export of 70–80% of its generation and, (iii) the US–Canada blackout of 14 August 2003. All systems are represented by notional parameters and topology which makes understanding easier.

6.2 Mumbai Network

Geography of Mumbai and its outskirts is shown in Fig. 6.1. Distribution network is shown in Fig. 6.2.

Mumbai is geographically an island, connected to the 400 kV network of MSETCL. 220 and 110 kV lines connect its subnetworks, viz. TATA (TECL) and ADANI (AEML) [1]. ICTs are installed for separation of these two subnetworks when found necessary. Blackout on Oct. 12, 2020 arose from discontinuance of import of about 1000 MW which comes on the 400 kV lines from PADGHE and TALEGAON (MSETCL) (Fig. 6.2). Mumbai has its own average generation of about 1350 MW and a load of 2600 MW. The ICT (bus# 6) at Kalwa has an RPUF (reverse power underfrequency relay) which operates at 47.9 Hz. Underfrequency relays are also located at the interconnecting transformers between Saki-A and Saki-T, and between Aarey and Borivli-M which separate Tata and AEML subnetworks. These

© The Author(s), under exclusive license to Springer Nature Singapore Pte Ltd. 2023   45
S. D. Varwandkar and M. V. Hariharan, *Fractal Tomography for Power Grids*,
SpringerBriefs in Computational Intelligence,
https://doi.org/10.1007/978-981-99-3443-0_6

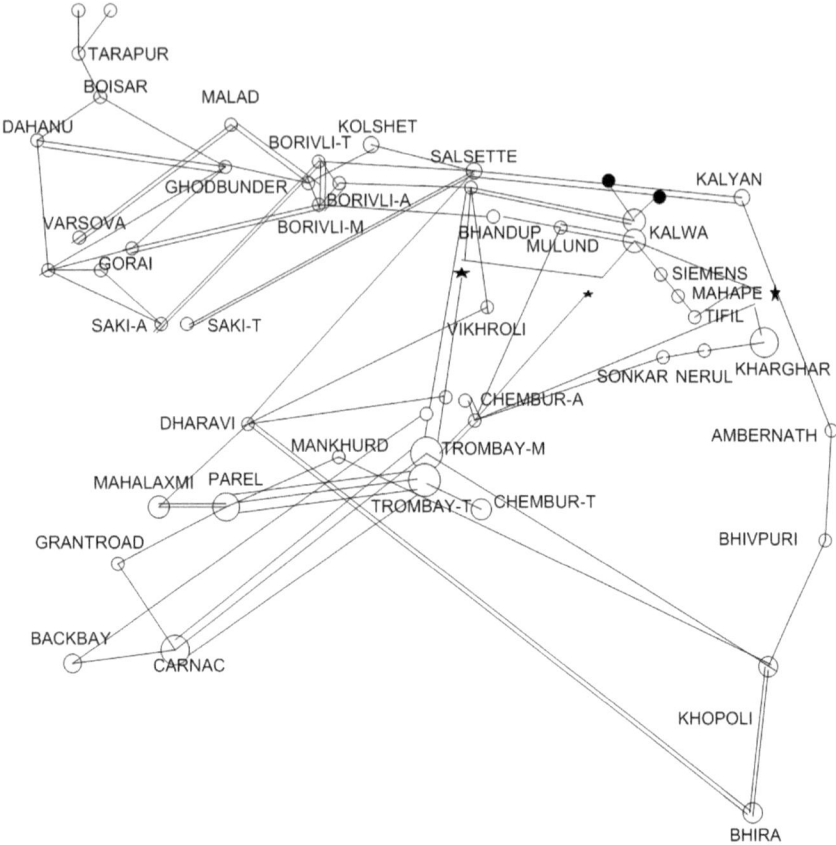

Fig. 6.1 Geographical network of Mumbai

operate at 48.5 Hz. Notional values for circuit parameters are used. Fractal tomography is used for determination of present and the future, which consist of line flows and voltages for various outaged stages, and preceding stages of the final blackout. Nominal data are given in Tables 6.1 and 6.2.

Following stages are considered [2]. Refer Fig. 6.2.

- Base case.
- Outage 1: PADGHE-KALWA (line 1–5): This outage transfers power to alternative import paths via TALEGAON-KHARGHAR-KALWA.
- Outage 2: Loss of line 3–5 (KHARGHAR-KALWA) blocking the import of 1023 MW from PADGHE and 387 MW from TALEGAON minus power supplied to NAVI MUMBAI.
- Outage 3: Line 4–5 (NAVI MUMBAI-KALWA), blocking the import 100%.
- Outage 4: Loss of line 24–31 (ICT between SAKINAKA-T and SAKINAKA-A (48.5 Hz UF)), and,

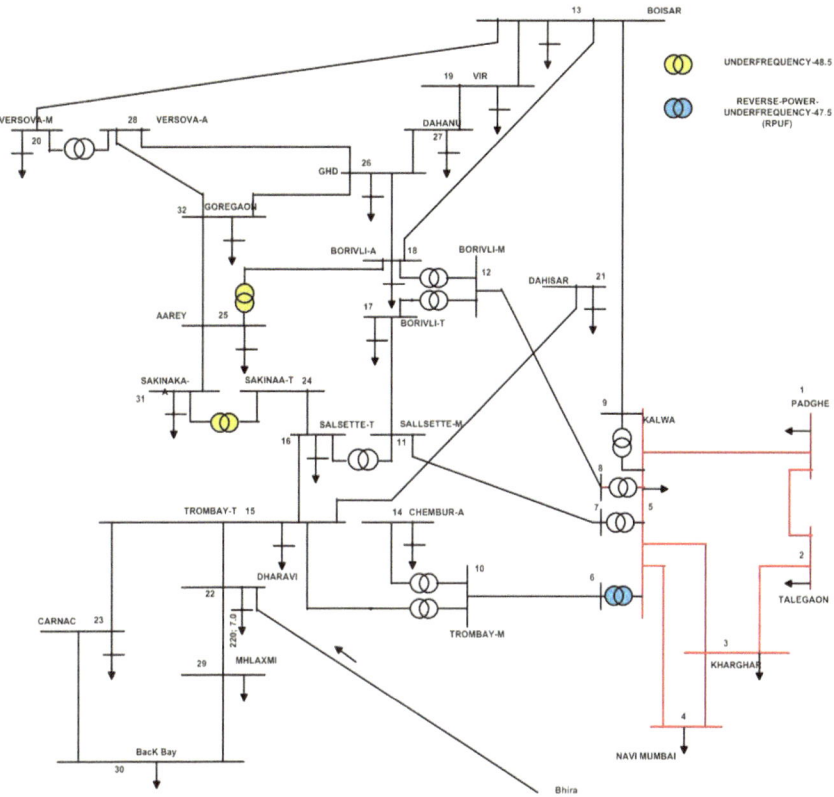

Fig. 6.2 Distribution network of Mumbai

- Outage 5: Line 18–35 (ICT between AAREY-A and BORIVLI (48.5 Hz UF)). It separates TECL and AEML distribution circuits.

Voltage tomographs are computed for (i) Base case, (ii) Outage 1, (iii) Outage 2, (iv) Outage 3, (v) Outage 4 and (vi) Outage 5. Only selective buses at SALSETTE (TECL), i.e. MULUND # 16, GOREGAON (AEML) # 32, VERSOVA (AEML) # 28, MAHALAXMI (TECL) # 29 and BACKBAY (TECL) # 30 are included in the tables. These are representative locations spanning the MUMBAI distribution network.

6.3 Salient Observations

Mumbai imports close to half of its requirement from Maharashtra State Electricity Transmission Limited (MSETCL). Available generating units within Mumbai are not kept on bar, if not scheduled. These are therefore not available as spinning reserves.

Table 6.1 Line data

From bus	To bus	R, pu	X, pu	$B/2$
1	2	0.0016	0.02	0.034
1	5	0.0005	0.006	0.0375
2	3	0.0015	0.002	0.03
3	4	0.0002	0.001	0.001
3	5	0.0001	0.001	0.001
4	5	0.0001	0.005	0.001
5	6	0.0001	0.002	0.001
5	7	0.0001	0.002	0.001
5	8	0.0001	0.002	0.001
5	9	0.0001	0.002	0.001
6	10	0.0004	0.005	0.001
7	11	0.0004	0.005	0.001
8	12	0.0004	0.005	0.001
9	13	0.0004	0.005	0.001
10	14	0.0001	0.002	0.001
10	15	0.0001	0.002	0.001
11	16	0.0001	0.002	0.001
11	17	0.0015	0.0015	0.001
12	17	0.0001	0.002	0.001
12	18	0.0001	0.002	0.001
13	18	0.001	0.01	0.001
13	19	0.0005	0.006	0.001
13	20	0.0025	0.025	0.001
15	16	0.001	0.01	0.001
15	21	0.0015	0.015	0.001
15	22	0.0005	0.005	0.001
15	23	0.0005	0.005	0.001
16	24	0.001	0.01	0.001
18	25	0.0001	0.002	0.001
18	26	0.001	0.002	0.001
19	27	0.0005	0.005	0.001
20	28	0.0001	0.002	0.001
22	29	0.0005	0.005	0.001
23	30	0.0005	0.005	0.001
24	31	0.0001	0.002	0.001
25	31	0.001	0.01	0.001

(continued)

Table 6.1 (continued)

From bus	To bus	R, pu	X, pu	$B/2$
25	32	0.001	0.01	0.001
26	27	0.001	0.01	0.001
26	28	0.005	0.05	0.001
26	32	0.005	0.05	0.001
28	32	0.005	0.05	0.001
29	30	0.0005	0.005	0.001
1	33	0.05	0.1	0.0001
2	34	0.05	0.1	0.0001
3	35	0.05	0.1	0.0001
4	36	0.05	0.1	0.0001
5	37	0.05	0.1	0.0001
6	38	0.05	0.1	0.0001
7	39	0.05	0.1	0.0001
8	40	0.05	0.1	0.0001
9	41	0.05	0.1	0.0001
10	42	0.05	0.1	0.0001
11	43	0.05	0.1	0.0001
12	44	0.05	0.1	0.0001
13	45	0.05	0.1	0.0001
14	46	0.05	0.1	0.0001
15	47	0.05	0.1	0.0001
16	48	0.05	0.1	0.0001
17	49	0.05	0.1	0.0001
18	50	0.05	0.1	0.0001
19	51	0.05	0.1	0.0001
20	52	0.05	0.1	0.0001
21	53	0.05	0.1	0.0001
22	54	0.05	0.1	0.0001
23	55	0.05	0.1	0.0001
24	56	0.05	0.1	0.0001
25	57	0.05	0.1	0.0001
26	58	0.05	0.1	0.0001
27	59	0.05	0.1	0.0001
28	60	0.05	0.1	0.0001
29	61	0.05	0.1	0.0001
30	62	0.05	0.1	0.0001
31	63	0.05	0.1	0.0001

(continued)

Table 6.1   (continued)

From bus	To bus	R, pu	X, pu	$B/2$
32	64	0.05	0.1	0.0001

Table 6.2   Generation and load data (base case)

220 kV bus	LV load bus	Load		Generation (220 kV)
		MW	MVAR	
1 (Padghe)	33	5	0	1023 (Padghe)
2 (Talegaon)	34	5	0	387 (Talegaon)
3	35	200	50	0
4	36	387	18	0
5	37	5	0	0
6	38	5	0	0
7	39	5	0	0
8	40	5	0	0
9	41	5	0	0
10	42	5	0	0
11	43	5	0	0
12	44	5	0	0
13	45	5	0	0
14	46	5	0	0
15 (Trombay)	47	88	10	864 (Trombay + Hydro)
16	48	80	10	0
17	49	90	10	0
18	50	130	20	0
19	51	30	0	0
20	52	5	0	0
21	53	110	10	0
22	54	90	10	0
23	55	310	70	0
24	56	300	50	0
25	57	87	10	0
26	58	75	10	0
27 (Dahanu)	59	115	10	485 (Dahanu)
28	60	115	10	0
29	61	280	30	0
30	62	185	20	0
31	63	200	70	0
32	64	170	60	0

Maintenance of transmission lines is crucial to ensure uninterrupted import. It is seen from Tables 6.3, 6.4 and 6.5 that the voltages are good until the outage of 4–5 which was the last 400 kV line connected to the network. This final outage isolates Mumbai region from the 400 kV network of MSETCL. Prospective voltages drop to almost 60% of the normal values (Table 6.6). Note that the tomographs gives voltage-information of even the islanded parts of the system. This is interesting since conventional power flow methods do not work after the network separates into more than one. The Jacobian is formed of partial derivative with respect to variables at all buses, which is not possible for disconnected networks.

Table 6.3 Voltage tomograph (base case)

Bus	PADGHE	TALEGAON	TECL	AEML	Net voltage
16 (Mulund)	0.41246	0.152449	0.347654	0.19133	1.1039
32 (Goregaon)	0.406044	0.150078	0.333766	0.196595	1.0865
28 (Versova)	0.410481	0.151717	0.336102	0.200793	1.0991
29 (Mahalaxmi)	0.396612	0.146591	0.361067	0.181852	1.0861
30 (Backbay)	0.396391	0.14651	0.360866	0.181751	1.0855

Table 6.4 Voltage tomograph (1–5 out)

Bus	PADGHE	TALEGAON	TECL	AEML	Net voltage
16 (Mulund)	0.383655	0.151285	0.34762	0.191309	1.079
32 (Goregaon)	0.377687	0.148932	0.333678	0.196624	1.0569
28 (Versova)	0.381814	0.150559	0.336001	0.200832	1.0692
29 (Mahalaxmi)	0.368914	0.145472	0.361224	0.181813	1.0574
30 (Backbay)	0.368708	0.145391	0.361022	0.181712	1.0568

Table 6.5 Voltage tomograph (1–5 and 3–5 out)

Bus	PADGHE	TALEGAON	TECL	AEML	Net voltage
16 (Mulund)	0.377019	0.148649	0.348026	0.191538	1.0652
32 (Goregaon)	0.371154	0.146337	0.334037	0.196885	1.0484
28 (Versova)	0.37521	0.147936	0.336357	0.201102	1.0606
29 (Mahalaxmi)	0.362533	0.142937	0.361748	0.18202	1.0492
30 (Backbay)	0.36233	0.142858	0.361546	0.181918	1.0487

Table 6.6 Voltage tomograph (1–5, 3–5 and 4–5 out—Mumbai islanded)

Bus	PADGHE	TALEGAON	TECL	AEML	Net voltage
16 (Mulund)	0	0	0.403538	0.223013	0.626551
32 (Goregaon)	0	0	0.388599	0.227918	0.616516
28 (Versova)	0	0	0.39149	0.232472	0.623962
29 (Mahalaxmi)	0	0	0.414874	0.212245	0.627119
30 (Backbay)	0	0	0.414642	0.212126	0.626768

6.4 Disturbances in Bhutan

Figure 6.3 shows western Bhutan [3] which generates substantial hydropower and exports 1000–1500 MW to the Indian network. A line diagram of Bhutan network is given in Fig. 6.4. Data are given in Tables 6.7, 6.8 and 6.9.

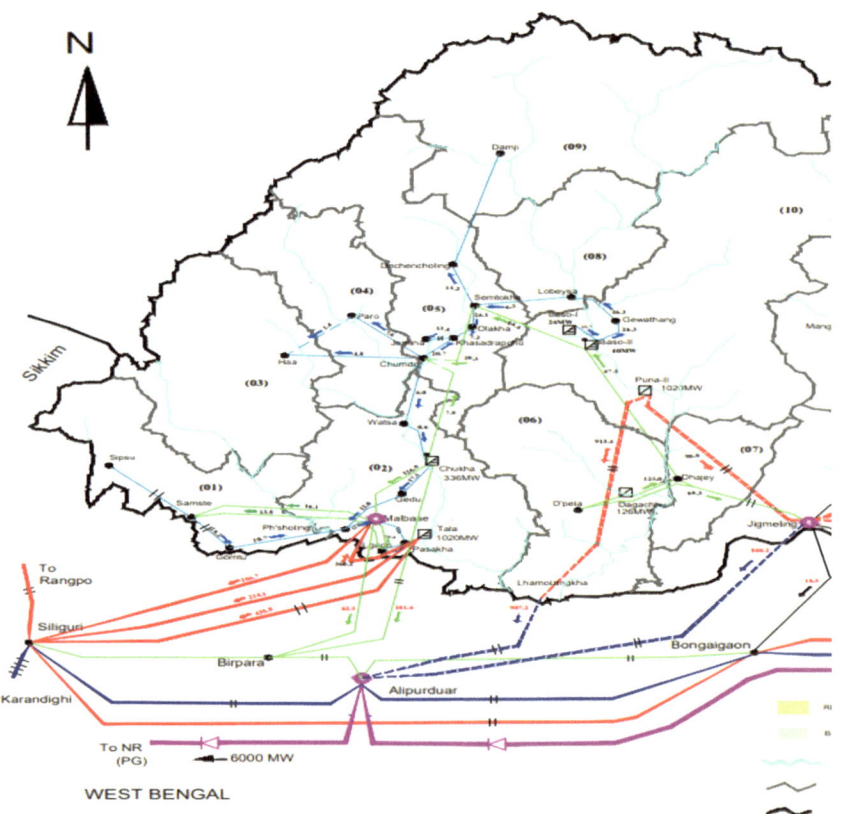

Fig. 6.3 Bhutan map

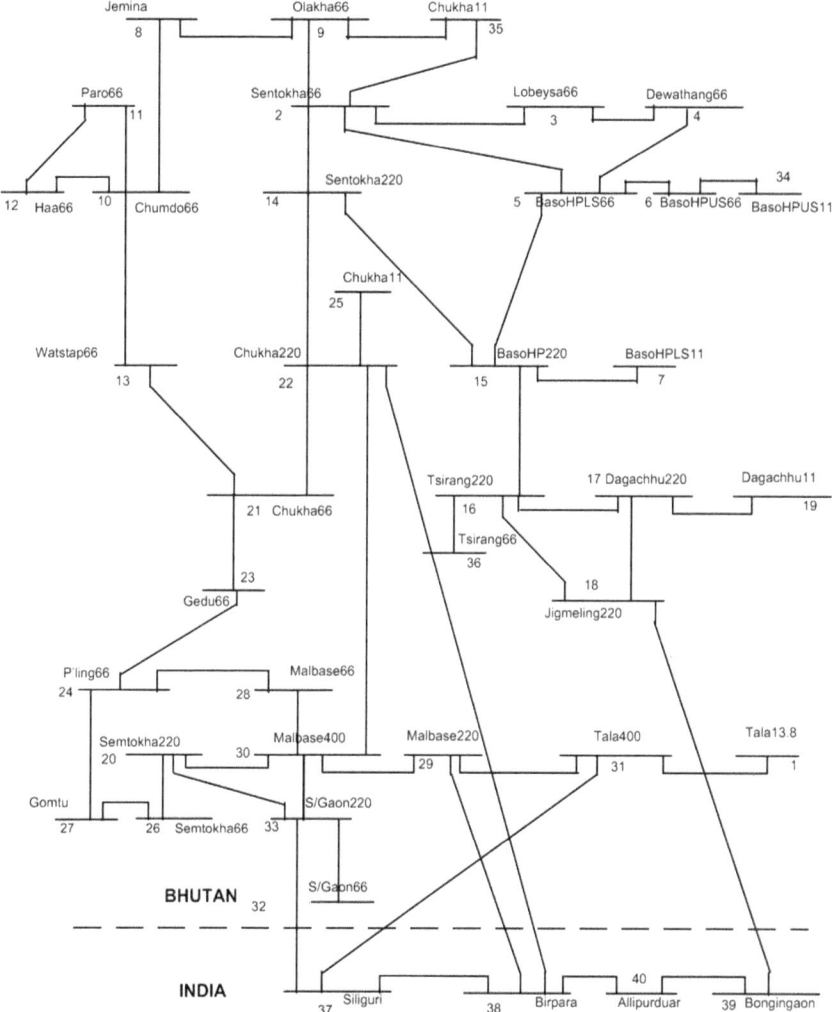

Fig. 6.4 Bhutan network

Bhutan load is 236.39 MW. Total generation is 1683.2 MW. Exports on Siliguri, Birpara and Boingangaon corridors are 1200, 100 and 148.8 MW respectively. There are 4 units of 92 MW each at CHUKHA (Tables 6.10, 6.11 and 6.12).

Problem statement: To analytically determine percentage change in voltages if 2 out of the 4 generators at Chukha are outaged simultaneously.

Solution: Voltage tomograph for the base case (with 4 generators) is given in Table 6.10.

Comments: Voltage tomographs are used to compute change in voltage due to outage of generators. In iterative power flows, this computation can be problematic

Table 6.7 Line data of Bhutan network

No	Fr.	To	R, pu	X, pu	$B/2$, pu	Tap		Fr.	To	R, pu	X, pu	$B/2$	Tap
1	1	31	0.00322	0.14496	0.001	1.05	22	2	14	0.00411	0.12353	0.001	1
2	31	29	0	0.005	0.145	1	23	2	35	0.043	0.101	0.001	1
3	30	29	0.00111	0.04998	0.001	1	24	2	3	0.091	0.216	0.003	1
4	33	30	0	0.001	0.002	1	25	14	15	0.006	0.037	0.063	1
5	33	32	0.00340	0.07831	0.001	1	26	15	7	0.00519	0.12989	0.001	1
6	20	30	0.006	0.033	0.057	1	27	15	16	0.007	0.038	0.066	1
7	22	30	0.004	0.025	0.042	1	28	16	36	0.00527	0.08443	0.001	1
8	26	20	0.00416	0.12503	0.001	1	29	16	18	0.004	0.025	0.043	1
9	27	26	0.055	0.131	0.002	1	30	16	17	0.003	0.017	0.028	1
10	24	27	0.1	0.238	0.003	1	31	17	18	0.008	0.045	0.078	1
11	23	24	0.062	0.148	0.002	1	32	19	17	0.00374	0.11994	0.001	1
12	21	23	0.076	0.181	0.003	1	33	5	15	0.00519	0.12989	0.001	1.02
13	22	21	0.00567	0.12487	0	1	34	5	6	0.011	0.027	0.001	1
14	33	20	0.006	0.034	0.058	1	35	34	6	0.00693	0.12480	0.001	1.02
15	21	13	0.081	0.194	0.003	1	36	9	2	0.006	0.015	0.001	1
16	13	10	0.056	0.134	0.002	1	37	3	4	0.081	0.193	0.003	1
17	10	11	0.089	0.213	0.003	1	38	4	5	0.006	0.013	0.001	1
18	10	12	0.124	0.296	0.004	1	39	24	28	0.033	0.08	0.001	1
19	10	8	0.044	0.104	0.002	1	40	28	30	0.00416	0.12503	0.001	1
20	8	9	0.066	0.156	0.002	1	41	25	22	0.00337	0.12495	0.001	1
21	22	14	0.008	0.044	0.076	1	–	–	–	–	–	–	–
Line data for last 4 (export) lines are assumed.							42	32	37	0.0011	0.050	0.002	1
							43	22	38	0.003	0.058	0.042	1
							44	29	38	0.0011	0.049	0.057	1
							45	31	37	0.0010	0.051	0.057	1
							46	18	39	0.0015	0.060	0.076	1

Table 6.8 Generator data

S. No	At bus	MW
1	1 TALA	1122
2	7 BASO-I	24.59
3	19 DAGACHHU	126.87
4	34 BASO-II	41.72
5	25 CHUKHA (92 × 4 generators)	368

Table 6.9 Load data

S. No	At bus	MW	MVAR
1	32	74	27
2	26	3.91	0.2
3	27	7.42	−6.7
4	24	6.34	2.22
5	23	1.69	0.32
6	21	1.15	0
7	13	0.58	0.06
8	10	0.89	0
9	11	7.22	1.27
10	12	8.7	0
11	8	1.92	0
12	9	7.6	1.8
13	2	10.14	0
14	35	5.53	1.36
15	3	7.22	1.18
16	34	0.17	0
17	36	2.38	0
18	17	0.3	0
19	19	0.3	0
20	6	0.89	0
21	34	0.88	0
22	4	12.415	−2.298
23	28	73.6	37.4
24	25	1.15	0
25	37	1100	30*
26	38	100	20*
27	39	148.8*	10*

* Denotes assumed values

Table 6.10 Voltage tomographs for selected buses: base case (4 generators)

Bus #	G1	G2	G3	G4	G5	Net voltage
11	0.636047	0.015541	0.066907	0.02237	0.23731	0.978175
12	0.634349	0.015499	0.066728	0.02231	0.236677	0.975563
13	0.654615	0.014027	0.060389	0.020133	0.243846	0.99301
14	0.660366	0.021243	0.091455	0.030382	0.24712	1.050566
15	0.657151	0.033869	0.145815	0.047395	0.245919	1.130149

Table 6.11 Voltage tomographs for selected buses: two generators outaged at Chukha

Bus #	G1	G2	G3	G4	G5	Net voltage
11	0.636047	0.015541	0.066907	0.02237	0.118655	0.85952
12	0.634349	0.015499	0.066728	0.02231	0.118338	0.857225
13	0.654615	0.014027	0.060389	0.020133	0.121923	0.871087
14	0.660366	0.021243	0.091455	0.030382	0.12356	0.927006
15	0.657151	0.033869	0.145815	0.047395	0.122959	1.007189

Table 6.12 Change in voltage due to outage of two generators at Chukha

Bus #	Voltage with 4 generators at Chukha	Voltage with 2 generators at Chukha	Percent change in voltage
11	0.978175	0.85952	12.13
12	0.975563	0.857225	12.13
13	0.99301	0.871087	12.27
14	1.050566	0.927006	11.76
15	1.130149	1.007189	10.87

since net imbalance, which is negative in this case, must be allocated to the slack bus. Voltage tomographs have no slack bus in formulation. It can therefore prove useful in planning outages for maintenance purposes.

6.5   US–Canada Blackout of 14 August 2003

The blackout in US–Canada region is explained in detail in this section. Symptoms of this blackout started at 01.31 pm on August 14, 2003, with the outage of nuclear units 4 and 5 situated at the south of the Lake Erie and bordering the northern borders of Ohio–Pennsylvania [4]. These units were feeding substantial power (and reactive power) to the state of Ohio. Keeping voltages normal became difficult. The load started drawing power from PJM and AEP, overloading lines in this region. Sequential outages of transmission lines on PJM-OHIO and AEP-OHIO corridors occurred shortly thereafter. Final collapse was triggered by outages of Argenta, Hampton-Pontiac, and Thetford-Jewel lines between Michigan and Detroit with several generators dropping in Michigan. These outages ultimately cut-off west Michigan from the east (Detroit). Loads in Detroit and OHIO started drawing power from Ontario via north of the Lake Erie, thereby reversing power flow on the IESO-Detroit lines. Earlier to the outage, this power came from the Michigan side. A tomographic study of these events is undertaken in this chapter.

6.5.1 The State of Ohio

Figure 6.5 shows OHIO in greater details. It shows locations of outaged transmission lines (corridors) [4] and helps ready visualization of the cascading sequence. The full US–Canada region as shown in Fig. 6.6.

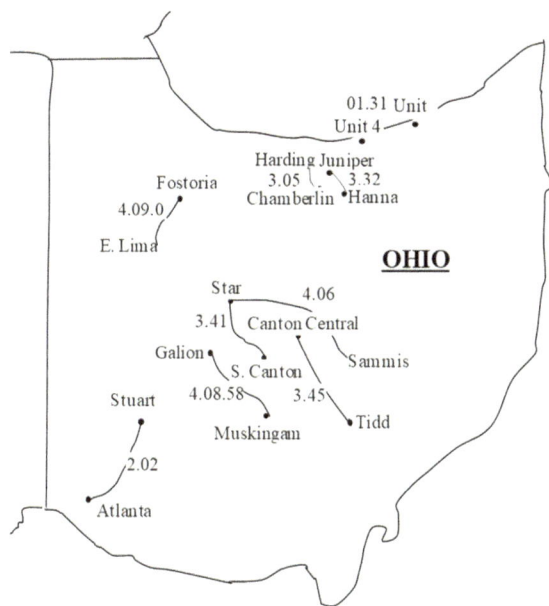

Fig. 6.5 Outaged lines in OHIO

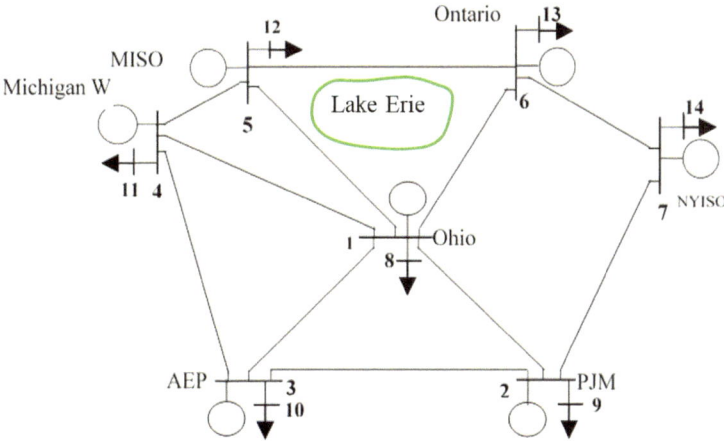

Fig. 6.6 Notional network of US–Canada

Data are given in Chap. 5. The generation schedules are kept unchanged.

6.6 Scenarios

Outage of Units 4 and 5 at 01.31 pm on the south of Lake Erie caused real and reactive power stress in OHIO which continued to operate normally until the outage of line 1–6 (Harding-Chamberlain and Jupiter-Hanna lines) at 03.05–3.32 pm. Next to go were lines 1–2 and 1–3—the outage of Canton Central-South Canton, Canton Central-Tidd, Galion-Muskingam and Star-Sammis lines representing the PJM-OHIO and the AEP-OHIO corridors. This was followed by outage of line 1–4 disrupting flow on the Lower Michigan-OHIO corridors. Subsequently lost are the lines 3–4 and 4–5 which constitute backbone in Michigan. The loads in Detroit and OHIO lost substantial power from Michigan. Outage of line 4–5 was the final blow with bulk generation drops in Michigan. Engineering judgement leads to conclusion that power from AEP to OHIO would be diverted to the path PJM to NYISO to IESO to Detroit to OHIO. This would cause many reversals and loop flows. Tomographic analysis quantifies these.

6.6.1 Tomographic Study

Quantitative analysis of the above scenario uses fractal tomographs for each stage. The six stages of cascading outages are as follows:

0. Base Case
1. Line 1–6 out
2. Line 1–6 and 1–2 out
3. Line 1–6, 1–2 and 1–3 out
4. Line 1–6, 1–2, 1–3 and 1–4 out
5. Line 1–6, 1–2, 1–3, 1–4 and 3–4 out
6. Line 1–6, 1–2, 1–3, 1–4, 3–4 and 4–5 out.

Tomographs for these stages are shown in Tables 6.13, 6.14, 6.15, 6.16, 6.17, 6.18 and 6.19. Although the tables appear nebulous, all fractals are not needed and can be selectively chosen and determined. Only those required in a particular study need to be chosen. Tomographs can address specific questions about the *past*, *present* and *future* of the grid (like ideal horoscopes). A few examples are given at the end. Tomographs have been calculated *off-line* for given outages (Tables 6.20 and 6.21).

Table 6.13 Lineflow tomograph for base case

Base case		G1	G2	G3	G4	G5	G6	G7
1	2	3069.604	−14,882.8	1836.62	3773.685	2085.78	5343.931	−961.449
1	3	1973.527	−643.138	−7770.7	−218.774	953.1328	3287.692	1272.171
1	4	819.7456	1782.475	−562.954	−3753.83	−194.038	725.1676	700.763
1	5	1443.57	4107.102	1044.329	−2440.58	−3128.47	−2059.89	13.15155
1	6	**1819.176**	**5336.968**	**3188.5**	**1059.865**	**−402.923**	**−9164.65**	**−2765.05**
2	3	−675.889	14,786.84	−11,005.5	−3901.69	−909.355	−1358.49	2502.924
2	7	479.1608	9843.174	2657.578	993.4568	151.5882	−1103.79	−11,792
3	4	−262.505	4373.028	7002.814	−7627.18	−1330.84	−1738.86	183.4682
4	5	−584.511	−340.215	2480.434	7529.239	−2688.59	−3834.63	−1716.55
5	6	367.5855	1214.004	2136.547	3581.573	2910.386	−7239.5	−2780.11
6	7	885.5876	−544.821	1591.877	1896.682	1134.118	5224.154	−8875.73

(Boldface represents pre-outage flow on the 1st outaged line)

Table 6.14 Lineflow tomograph with line 1–6 out (stage 1)

First stage		G1	G2	G3	G4	G5	G6	G7
1	2	3453.337	−13,782.5	2518.71	4005.771	1985.347	3062.55	−1516.42
1	3	2245.175	153.4779	−7314.49	−60.8577	885.8163	1730.726	845.2693
1	4	1034.899	2420.628	−184.216	−3636.98	−240.079	−328.133	365.2458
1	5	2357.889	6828.437	2663.941	−1906.41	−3305.07	−6026.3	−1366.33
1	6	0	0	0	0	0	0	0
2	3	−728.734	14,587.66	−11,101.7	−3939.77	−891.708	−964.651	2556.408
2	7	844.1099	10,999.77	3327.876	1216.268	67.5836	−2747.17	−12,182.6
3	4	−90.0044	4897.186	7317.798	−7534.11	−1361.84	−2377.72	−86.1462
4	5	−210.849	778.8324	3143.676	7765.105	−2762.05	−5270.31	−2269.13
5	6	1613.937	4991.814	4381.97	4350.828	2622.264	−12,998.9	−4747.05
6	7	456.4269	−1737.03	841.2328	1638.427	1239.708	7400.333	−8368.82
1	8	424.3501	2190	1158.027	799.2373	336.9929	780.5766	836.1173

Table 6.15 Lineflow tomograph with lines 1–6, 1–2 out (stage 2)

Second stage		G1	G2	G3	G4	G5	G6	G7
1	2	0	0	0	0	0	0	0
1	3	3990.986	−6555.18	−6096.11	2027.464	1928.322	3367.646	45.71778
1	4	1554.85	306.1191	201.3829	−3082.56	63.03893	139.8093	133.3001
1	5	3374.482	2540.342	3454.916	−728.8	−2755.1	−5201.29	−1793.99
1	6	0	0	0	0	0	0	0
2	3	−2744.33	24,463.19	−12,639.9	−6362.71	−2124.44	−3017.17	3683.6
2	7	188.1459	13,629.23	2801.782	436.2148	−287.044	−3210.37	−11,961.4
3	4	−729.937	7406.157	6798.536	−8209.64	−1720.83	−2980.78	231.6509
4	5	−472.897	1773.749	2954.018	7412.319	−2914.68	−5550.1	−2124.33
5	6	2234.785	2095.733	4920.597	5181.332	3041.344	−12,449.2	−5007.82
6	7	1004.17	−3979.99	1290.425	2328.16	1574.012	7944.199	−8578.04
1	8	509.8412	1854.36	1219.905	891.9485	381.8673	846.915	807.4843
1	8	509.8412	1854.36	1219.905	891.9485	381.8673	846.915	807.4843

6.6.2 Past, Present and the Future

Query 1. How much power is carried by line 1–6 before the 1st outage?

Answer: "−928.07 MW" ('minus' means *from 6 to 1*). This is sum of the fractals in 5th row for line 1–6 in Table 6.13. This is also shown in the last column of Table 6.20 and highlighted in the Table 6.21.

Table 6.16 Lineflow tomograph with lines 1–6, 1–2, 1–3 out (stage 3)

Third stage		G1	G2	G3	G4	G5	G6	G7
1	2	0	0	0	0	0	0	0
1	3	0	0	0	0	0	0	0
1	4	2963.914	−2063.09	−1935.56	−2415.06	785.6144	1419.081	150.2896
1	5	5574.936	−1214.02	−60.7484	428.1249	−1724.54	−3391.64	−1770.29
1	6	0	0	0	0	0	0	0
2	3	−1665.95	22,606.02	−13,954.3	−5774.63	−1604.93	−2185	3693.451
2	7	−357.517	14,727.71	3787.073	104.8996	−555.723	−3596.97	−11,962.8
3	4	−2836.8	12,186.87	11,202.65	−9333.29	−2818.31	−5080.82	199.9514
4	5	−1653.4	4032.205	4931.171	6704.242	−3506.37	−6681.26	−2141.04
5	6	2847.622	733.3537	3509.193	5653.524	3441.402	−11,846.7	−5002.43
6	7	1500.745	−4946.49	254.8183	2668.565	1859.189	8465.14	−8576.26
1	8	700.5754	1638.555	998.155	993.4677	469.4638	986.2772	809.9981
1	8	700.5754	1638.555	998.155	993.4677	469.4638	986.2772	809.9981
2	9	1011.736	13,314.13	5083.599	2834.864	1080.326	2890.988	4134.653

Table 6.17 Lineflow tomograph with lines 1–6, 1–2, 1–3 and 1–4 out (stage 4)

Fourth stage		G1	G2	G3	G4	G5	G6	G7
1	2	0	0	0	0	0	0	0
1	3	0	0	0	0	0	0	0
1	4	0	0	0	0	0	0	0
1	5	8173.891	−3133.41	−1853.66	−1773.21	−1038.18	−2130.61	−1628.91
1	6	0	0	0	0	0	0	0
2	3	−1256.22	22,308.13	−14,182.4	−6007.52	−1496.69	−2006.61	3711.224
2	7	−509.844	14,897.61	3955.545	299.8781	−607.439	−3670.95	−11,970.9
3	4	−2234.91	11,851.29	10,852.52	−9647.48	−2662.72	−4829.17	221.3149
4	5	−3384.15	5781.681	6570.057	8746.756	−4031.95	−7700.42	−2265.97
5	6	2960.352	495.3879	3243.618	5302.336	3527.026	−11,715.5	−4988.57
6	7	1608.639	−5100.71	72.35125	2426.347	1916.567	8571.45	−8566.73
1	8	883.0547	1566.706	926.8319	886.6065	519.0907	1065.304	814.4558
1	8	883.0547	1566.706	926.8319	886.6065	519.0907	1065.304	814.4558
2	9	883.0324	13,378.13	5113.436	2853.82	1052.064	2838.778	4129.852
2	9	883.0324	13,378.13	5113.436	2853.82	1052.064	2838.778	4129.852

Table 6.18 Lineflow tomograph with lines 1–6, 1–2, 1–3, 1–4 and 3–4 out (stage 5)

Fifth stage		G1	G2	G3	G4	G5	G6	G7
1	2	0	0	0	0	0	0	0
1	3	0	0	0	0	0	0	0
1	4	0	0	0	0	0	0	0
1	5	7731.722	−1423.32	−578.071	−3014.7	−1519.21	−3199.21	−1595.78
1	6	0	0	0	0	0	0	0
2	3	290.2156	13,502.85	−20,493.5	657.2891	331.2309	1226.828	3530.421
2	7	−978.1	18,454	7494.931	−2215.23	−1116.33	−4134.72	−11,898.4
3	4	0	0	0	0	0	0	0
4	5	−2029.12	−2169.71	−881.21	15,655.66	−2315.89	−4876.87	−2432.6
5	6	4150.539	−4555.95	−1850.36	9400.267	4737.123	−10,240.4	−5107.96
6	7	2250.676	−7709.97	−3131.34	5097.4	2568.758	9514.284	−8644.13
1	8	1104.139	711.6621	289.0353	1507.35	759.6062	1599.605	797.8895
1	8	1104.139	711.6621	289.0353	1507.35	759.6062	1599.605	797.8895
2	9	343.942	16,002.58	6499.306	778.9704	392.5504	1453.946	4183.994
2	9	343.942	16,002.58	6499.306	778.9704	392.5504	1453.946	4183.994
3	10	141.1441	6567.007	5007.31	319.6676	161.0916	596.6587	1716.994

Table 6.19 Lineflow tomograph with lines 1–6, 1–2, 1–3, 1–4, 3–4 and 4–5 out (stage 6)

Sixth stage		G1	G2	G3	G4	G5	G6	G7
1	2	0	0	0	0	0	0	0
1	3	0	0	0	0	0	0	0
1	4	0	0	0	0	0	0	0
1	5	7195.379	−2170.72	−891.188	0	−2015.36	−4218.1	−2264.59
1	6	0	0	0	0	0	0	0
2	3	386.9481	13,394.25	−20,760.6	0	439.4047	1498.021	3677.459
2	7	−1304.11	18,820.02	7726.564	0	−1480.9	−5048.71	−12,394
3	4	0	0	0	0	0	0	0
4	5	0	0	0	0	0	0	0
5	6	5533.966	−3571.43	−1466.25	0	6284.178	−6939.94	−3725.88
6	7	3000.854	−7835.88	−3217.02	0	3407.665	11,617.44	−8174.76
1	8	1372.31	1085.358	445.5941	0	1007.679	2109.051	1132.297
1	8	1372.31	1085.358	445.5941	0	1007.679	2109.051	1132.297
2	9	458.5822	15,873.87	6517.022	0	520.7499	1775.344	4358.252
2	9	458.5822	15,873.87	6517.022	0	520.7499	1775.344	4358.252
3	10	188.1893	6514.19	4893.066	0	213.7011	728.5514	1788.504
3	10	188.1893	6514.19	4893.066	0	213.7011	728.5514	1788.504

Table 6.20 Row-sums (net lineflows) in base-case (a column added to Table 6.13)

Line		G1	G2	G3	G4	G5	G6	G7	Sum
1	2	3069.604	−14882.8	1836.62	3773.685	2085.78	5343.931	−961.449	**265.35**
1	3	1973.527	−643.138	−7770.7	−218.774	953.1328	3287.692	1272.171	**−1146.1**
1	4	819.7456	1782.475	−562.954	−3753.83	−194.038	725.1676	700.763	**−482.67**
1	5	1443.57	4107.102	1044.329	−2440.58	−3128.47	−2059.89	13.15155	**−1020.8**
1	6	1819.176	5336.968	3188.5	1059.865	−402.923	−9164.65	−2765.05	**−928.07**
2	3	−675.889	14786.84	−11005.5	−3901.69	−909.355	−1358.49	2502.924	**−561.16**
2	7	479.1608	9843.174	2657.578	993.4568	151.5882	−1103.79	−11792	**1229.2**
3	4	−262.505	4373.028	7002.814	−7627.18	−1330.84	−1738.86	183.4682	**599.93**
4	5	−584.511	−340.215	2480.434	7529.239	−2688.59	−3834.63	−1716.55	**845.17**
5	6	367.5855	1214.004	2136.547	3581.573	2910.386	−7239.5	−2780.11	**190.49**
6	7	885.5876	−544.821	1591.877	1896.682	1134.118	5224.154	−8875.73	**1311.9**

(Boldface denotes the sum of elements in the row)

Table 6.21 Row sums (net flows) of the five stages

Line		Base case	1st stage	2nd stage	3rd stage	4th stage	5th stage	6th stage
1	2	265	−273.25	0.0	0.0	0.0	0.0	0.0
1	3	**−1146.1**	**−1514.9**	−1291.2	0.0	0.0	0.0	0.0
1	4	−482.67	−568.64	−684.06	−1094.8	0.0	0.0	0.0
1	5	−1020.8	−753.84	−1109.4	−2158.2	−3384.1	−3598.6	−4364.6
1	6	**−928.07**	0.0	0.0	0.0	0.0	0.0	0.0
2	3	−561.16	−482.47	1258.2	1114.7	1069.9	−954.71	−1364.5
2	7	**1229.2**	1525.8	1596.5	2146.7	2393.9	5606.1	6318.9
3	4	599.93	765.17	795.15	3520.3	3550.8	0.0	0.0
4	5	845.17	1175.3	1078.1	1685.5	3716	950.26	0.0
5	6	**190.49**	214.87	16.752	−664	−1175.3	−3466.8	−3885.4
6	7	1311.9	1470.3	1582.9	1225.7	927.91	−54.317	−1201.7

(Boldface relate to the question-answers in Sect. 6.6.2)

Query 2: Which line carries maximum power? What is the value?

Answer: "Line 2–7". Row-sum for 2–7 in Table 6.13 is 1229.2 MW which is seen in the last column of Table 6.20 and highlighted in Table 6.21.

Query 3: Which line carries the minimum power before 1st outage?

Answer: "Line 5–6". It carries 190.49 MW (Table 6.13). Also, see last column of 6.20 and the highlight in Table 6.21.

Query 4: How much change (from the base case) will occur in, say, line 1–3 due to outage of line 1–6?

Answer: "368.8 MW." This is obtained from the sum of row 2 of stage 1 in Table 6.13 and that of row 2 of stage 2 in Table 6.14. The flow changes from -1146.1 to -1514.9 MW which is highlighted in Table 6.21.

There are many other queries that can be answered using tomograph details.

References

1. *Maharashtra Electricity Regulatory Commission.*: Transmission License No. 1 of 2014 (Second Amendment) Available: www.merc.gov.in
2. *Maharashtra Electricity Regulatory Commission.*: Case No. 202 of 2020: Suo Motu Proceeding in the matter of Grid Failure in the Mumbai Metropolitan Region on 12 October 2020 at 10.02 Hrs.: Available: www.merc.gov.in
3. Department of hydropower and power systems, Ministry of economic affairs, Royal Government of Bhutan, Thimpu, *National transmission grid master plan (NTGMP) of Bhutan-2018*, June 2018 [Available on the net]
4. Final report on the August 14, 2003 Blackout in the United States and Canada: Causes and Recommendations, April 2004. [Available on the net]

Chapter 7
Frequency Rendezvous

7.1 Introduction

Significant frequency deviations occur in power systems during disturbances giving rise to synchronizing power flows. After an outage, the steady state is reached at a new frequency which we call *frequency of rendezvous* (FR). This new equilibrium frequency is generally different from the original frequency. Disturbance dynamics is the travel of frequency from pre-outage equilibrium value to the post-outage equilibrium value, and is termed *frequency rendezvous dynamics* (FRD). Concept of frequency equilibrium is different from that of the power angle variation which *evolves from* an initial condition; the end point is reached at t = infinity. In FRD, generators *chase* a *final* frequency of rendezvous which takes finite time. This is better explained by an example. *All sprinters start from given set-points, but their dynamics consists in their "chasing" a target.* The same happens in *FRD,* but the target frequency moves in accordance with a weighted average the frequencies of the generators chasing it. The chase continues until frequencies of all generators reach the frequency of rendezvous. We derive expression for the *frequency of rendezvous* (FR). Lineflows and voltages in the network can be calculated from fractal tomographs computed at FR. Use of RX representation of circuit elements has an implicit assumption is that '*R*' is independent of frequency. For GB representation of the same element, '*G*' is considered independent of frequency. These two assumptions are not mathematically compliant with each other since '*G*' calculated from R and X is a function of frequency. Nonetheless, an electrical device may be represented by a series RX or a parallel GB model. Results of this chapter should be interpreted with these representational limitations in mind.

© The Author(s), under exclusive license to Springer Nature Singapore Pte Ltd. 2023   65
S. D. Varwandkar and M. V. Hariharan, *Fractal Tomography for Power Grids*,
SpringerBriefs in Computational Intelligence,
https://doi.org/10.1007/978-981-99-3443-0_7

7.2 Frequency Rendezvous Dynamics

We employ *one-machine-RL* (OMRL) model termed as a *module* [1]. A series '*R-L*' element is the driving point parameter of the network at the generator bus and is calculated from impedance parameter matrix of the system. Q is obtained from an operational multimachine network at nominal frequency and kept constant at that value. P of the source and Q are related as follows.

$$P = I^2 R \tag{7.1}$$

And,

$$Q = I^2 \lambda L \tag{7.2}$$

L can be calculated at nominal frequency from $X = \omega^{\text{nom}} L$ given as data. Frequency in the conceptual one-machine module is then given by,

$$\lambda = \frac{QR}{PL}. \tag{7.3}$$

This conceptual frequency is called *characteristic generator frequency* (CGF) of *i*th generator. Steady state value is λ_{si} and the dynamic, λ_i. Its variation with P is hyperbolic which is shown in Fig. 7.1. The two curves represent prefault and post-fault conditions.

For a frequency-change $\Delta\lambda$ during Δt we can write,

$$M_i(\Delta\lambda_i)|_{t=0} = M_i \frac{d\lambda_i}{dt}\bigg|_{t=0-} \Delta t. \tag{7.4}$$

Let the driving-point parameter network change from R_0, L_0 to R_s, L_s as a result of some outage. The operating point travels from '1' to '2', on the post-fault curve.

Fig. 7.1 Variation of network-frequency in the conceptual one-generator system

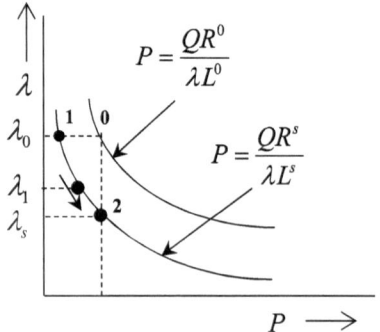

Points on this curve denote frequency which moves in discrete steps along the post-fault curve as per (7.5),

$$M_i \left(\frac{d\lambda_i}{dt} \right) = \text{Turbine input-network power} \tag{7.5}$$

Or,

$$M_i \left(\frac{d\lambda_i}{dt} \right) = \frac{Q_i R_{ii}^0}{\lambda_i L_{ii}^0} - \frac{Q R_{ii}^s}{\lambda_s L_{ii}^s} \tag{7.6}$$

With a little manipulation, the FRD equation can be reduced to,

$$\left(\frac{d\lambda_i}{dt} \right) = -\frac{P_i}{M_i} \left(1 - \frac{\lambda_{si}}{\lambda_i} \right); \lambda_{si} = \frac{Q_i R_{ii}^0}{P_i L_{ii}^0} \tag{7.7}$$

Negative sign indicates that *fall* in frequency is taken as positive. Above relations are for one-generator conceptual modules. With many generators, we have as many modules and as many *characteristic generator frequencies*, λ_{s1}, λ_{s2}, ...

7.3 Multigenerator Systems

In multigenerator systems with all generators connected to the network, interactions occur between generators through the network. One generator pulls and the other pushes by virtue of synchronizing powers. With system finally settling at ω_s, the interactions in case of a two-machine system can be expressed in FRD equation of generator 1 as

$$\text{Interaction } 1 = f_1(\lambda_1 - \omega_s) \tag{7.8}$$

And, in FRD equation of generator 2 as

$$\text{Interaction } 2 = f_2(\lambda_2 - \omega_s) \tag{7.9}$$

Interactions are mutual, and sum of all interactions is zero in steady state. For two-machine system,

$$f_1(\lambda_1 - \omega_s) + f_2(\lambda_2 - \omega_s) = 0 \tag{7.10}$$

FRD equations for actual machine frequencies ω_i can be written as [1],

$$\frac{d\omega_1}{dt} = \frac{P_1}{M_1} \left(1 - \frac{\lambda_{s1}}{\omega_1} \right) + \frac{P_1}{M_1} \left(\frac{\lambda_{s1} - \omega_s}{\omega_1} \right) \tag{7.11}$$

$$\frac{d\omega_2}{dt} = \frac{P_2}{M_2}\left(1 - \frac{\lambda_{s2}}{\omega_2}\right) + \frac{P_2}{M_2}\left(\frac{\lambda_{s2} - \omega_s}{\omega_2}\right) \tag{7.12}$$

First term on the RHS is 'self-dynamics' of generator, and the second term is 'interactive dynamics'.

Frequency stabilizes in steady state when,

$$\left.\frac{d\omega_1}{dt}\right|_{\omega_1=\omega_s} = \left.\frac{d\omega_2}{dt}\right|_{\omega_2=\omega_s} = 0 \tag{7.13}$$

At this frequency, *sum* of interactions is zero, that is,

$$\frac{P_1}{M_1}\left(\frac{\lambda_{s1} - \omega_s}{\omega_1(=\omega_s)}\right) + \frac{P_2}{M_2}\left(\frac{\lambda_{s2} - \omega_s}{\omega_2(=\omega_s)}\right) = 0 \tag{7.14}$$

This condition gives,

$$\omega_s = \frac{\frac{P_1}{M_1}\lambda_{s1} + \frac{P_2}{M_2}\lambda_{s2}}{\frac{P_1}{M_1} + \frac{P_2}{M_2}} \tag{7.15}$$

In steady state,

(i) derivatives of individual generator frequencies become zero and
(ii) interactions stabilize at $\frac{P_1}{M_1}\left(\frac{\lambda_{s1}-\omega_s}{\omega_s}\right)$ and $\frac{P_2}{M_2}\left(\frac{\lambda_{s2}-\omega_s}{\omega_s}\right)$.

Frequency ω_s can be viewed as *frequency of rendezvous* (FR) of generators. FR can be theoretically defined for dynamic conditions using (7.15) as,

$$\omega_s(t) = \frac{\frac{P_1}{M_1}\omega_1(t) + \frac{P_2}{M_2}\omega_2(t)}{\frac{P_1}{M_1} + \frac{P_2}{M_2}} \tag{7.16}$$

Replacement of λ's by ω's is justified since all interactions, being internal, sum to zero at all times. For multigenerator systems, (7.16) can be written as,

$$\omega_s = k_1\lambda_{s1} + k_2\lambda_{s2} + k_3\lambda_{s3} + \cdots + k_{ng}\lambda_{s,ng} \tag{7.17}$$

In (7.17),

$$k_i = \frac{\frac{P_1}{M_1}}{\sum_{i=1}^{ng}\frac{P_i}{M_i}} \tag{7.18}$$

FRD fractals are,

$$k_1 \lambda_{s1} = \frac{\frac{P_1}{M_1} \lambda_{s1}}{\sum_{i=1}^{ng} \frac{P_i}{M_i}}, \quad k_2 \lambda_{s2} = \frac{\frac{P_2}{M_2} \lambda_{s2}}{\sum_{i=1}^{ng} \frac{P_i}{M_i}}, \ldots, \text{ and so on.} \tag{7.19}$$

FRD behaviour for machine i contains a synchronizing term in (7.11), (7.12),

$$f_i = \frac{P_i \, \lambda_{si}}{M_i \, \omega_i} \tag{7.20}$$

The term appears in Eqs. (7.11) and (7.12) two times with opposite signs and does not affect the numerical value of ω_i. Interestingly, (7.11) and (7.12) can be written in terms of the rendezvous (system) frequency alone, i.e.,

$$\frac{d\omega_1}{dt} = \frac{P_1}{M_1} \left(1 - \frac{\omega_s}{\omega_1} \right) \tag{7.21}$$

And,

$$\frac{d\omega_2}{dt} = \frac{P_2}{M_2} \left(1 - \frac{\omega_s}{\omega_2} \right) \tag{7.22}$$

For multimachine systems,

$$\frac{d\omega_i}{dt} = \frac{P_i}{M_i} \left(1 - \frac{\omega_s}{\omega_i} \right); i = 1, 2, \ldots \tag{7.23}$$

Since the synchronizing terms of (7.11), (7.12) have got cancelled in (7.21), (7.22), we lose information about the internal dynamics of synchronizing behaviour.

7.4 Example 1

Consider a load fed by a two transmission lines as shown in Fig. 7.2. Inductance values are in pu Henry (nominal-reactance/314).

With 2 lines, i.e. $L_{line}^0 = \frac{0.24}{314} = 0.00075$.

With 1 line, i.e. $L_{line}^s = \frac{0.48}{314} = 0.0015$

$$L_{load}^0 = L_{load}^s = \frac{0.56}{314} = 0.0018;$$

New steady-state frequency is,

$$\omega_s = \omega_0 \frac{R^s L^0}{L^s R^0} = 256.0308 \text{ rad/s} = 40.77 \text{ Hz}$$

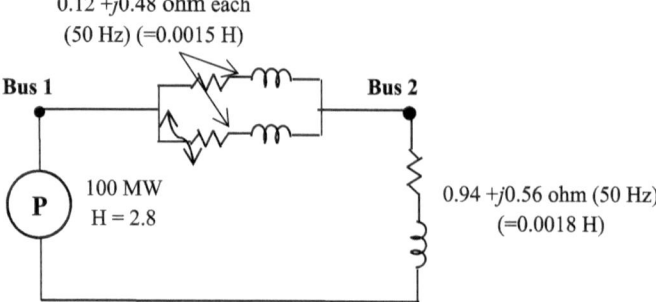

Fig. 7.2 Example 1

Let the generator power and network power in prefault steady state be, $P_m = P_n = 100$ MW. $H = 2$. Then the dynamic equation is,

$$\frac{2GH}{\omega^0}\left(\frac{d\omega}{dt}\right) = -100 * \left(1 - \frac{256.0308}{\omega}\right); \omega = \omega_0 \text{ at } t = 0.$$

In per unit, $H = 2.8$ MW s/MVA; 100 MVA base.

$$\frac{d\omega}{dt} = -\frac{\pi f^0}{2.8}\left(1 - \frac{256.0308}{\omega}\right); \omega = \omega_0 \text{ at } t = 0$$

At $\omega = \omega_s$, derivative is zero, i.e. $\frac{d\omega}{dt} = 0$. Time response is shown in Fig. 7.3. The equations are solved step-by-step.

7.5 Example 2

This is a typical system with medium and small generators connected to a large grid. Network for this example is shown in Fig. 7.4. There are three generators, (1) 50 MW, (2) 2.5 MW, both connected to bus 1 and (3) 250 MW to bus 2. The inertia (H) are 2.0, 1.0 and 3.5, in that order, on own ratings. Frequency rendezvous dynamics after outage of one of the lines is required to be studied. Initial operational system frequency is ω_0. Following initial computation is required.

$$Z_{net}^0 = \begin{bmatrix} 0.23548 + j0.155 & 0.23276 + j0.14699 \\ 0.23276 + j0.14699 & 0.23626 + j0.16248 \end{bmatrix};$$

$$Z_{net}^s = \begin{bmatrix} 0.23664 + j0.1577 & 0.23109 + j0.14187 \\ 0.23109 + j0.14187 & 0,23837 + j0.17231 \end{bmatrix}$$

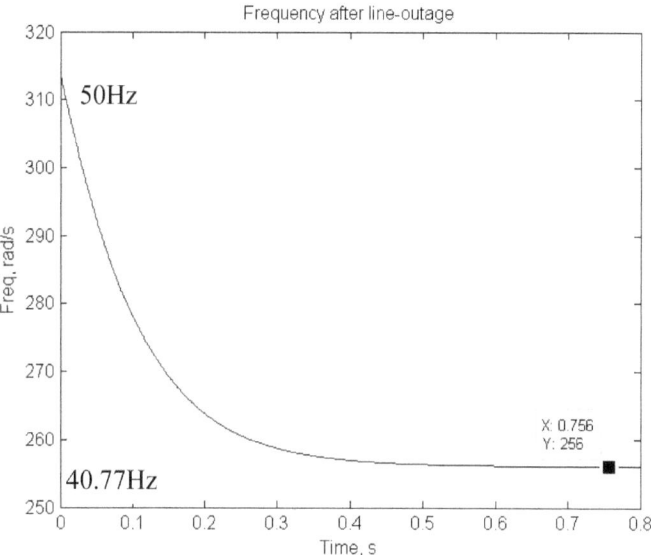

Fig. 7.3 Frequency response

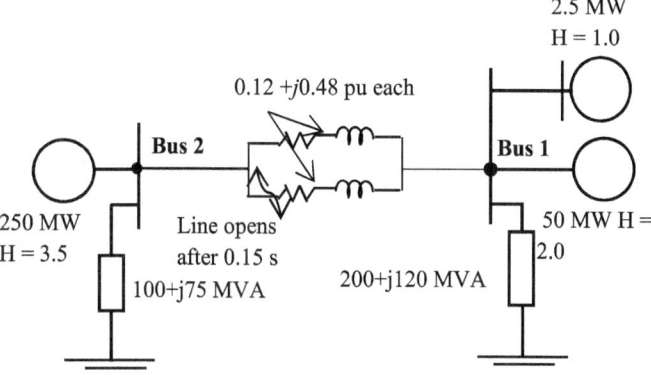

Fig. 7.4 Example 2

Initial (nominal) frequency is 314 rad/s (50 Hz). Characteristic frequencies of generators for outaged network are,

$$\lambda_{si} = \frac{R^s_{ii} L^0_{ii}}{L^s_{ii} R^0_{ii}} \omega_0$$

These are found to be 310.43 rad/s (49.40 Hz), 310.43 (49.4 Hz) and 298.89 (47.57 Hz) for generators 1, 2 and 3, respectively. Frequency of rendezvous is 49.112 Hz. FRD fractals are computed as shown below.

$$\frac{P_1}{M_1} = \frac{50}{\left(\frac{2*50*2.0}{314}\right)}; \quad \frac{P_2}{M_2} = \frac{2.5}{\left(\frac{2*2.5*1.0}{314}\right)} =; \quad \frac{P_3}{M_3} = \frac{250}{\left(\frac{2*250*3.5}{314}\right)}$$

That is,

$$\frac{P_1}{M_1} = 78.54; \quad \frac{P_2}{M_2} = 157.08; \quad \frac{P_3}{M_3} = 44.88$$

$$\sum_{i=1}^{3} \frac{P_i}{M_i} = 280.5$$

FRD fractals for generator 1, 2 and 3 are 86.92, 173.84 and 47.822, respectively [use (7.19)]. It is high for low-inertia machine. Motive force for frequency is $M\omega$. Therefore, *force-of-frequency* fractals can also be obtained by transferring the M-term in each FRD-fractal to LHS. *Force-of-frequency* (denoted by f_f) fractals are then,

Generator	1	2	3
f_v	86.92	173.08	47.822
f_f	50	2.5	250

Both, f_v and f_f can be plotted on a pie chart. Force of frequency fractals are shown in Fig. 7.5.

Steady-state frequency of rendezvous (FR) is 49.112 Hz. FR varies slowly with time. Its fractals are dynamic in nature and are shown in Fig. 7.6 for a fault on line 2 cleared after 0.15 s. Low-frequency oscillations occur about the FR as shown in Fig. 7.6. These are usually known as inter-area oscillations. The value of 49.112 Hz shown in Fig. 7.6 is the post-fault steady-state value. Time variation of FR is shown in Fig. 7.7. Inductances are constants, but reactances vary with FR. Impedances

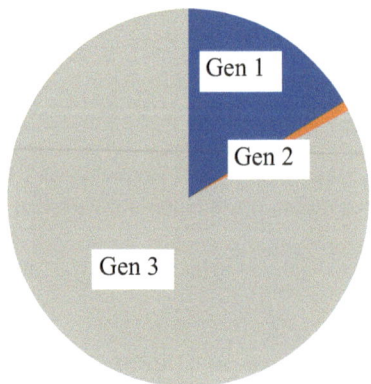

Fig. 7.5 Force-of-frequency fractals

change for large disturbances. Lineflows and voltages also turn dynamic. Fractal approach can do editing and obtain dynamic lineflows and voltages.

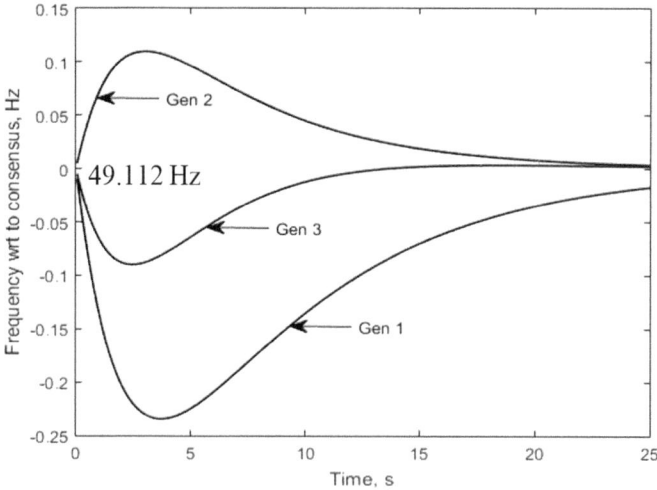

Fig. 7.6 Frequency deviation with respect to frequency of rendezvous

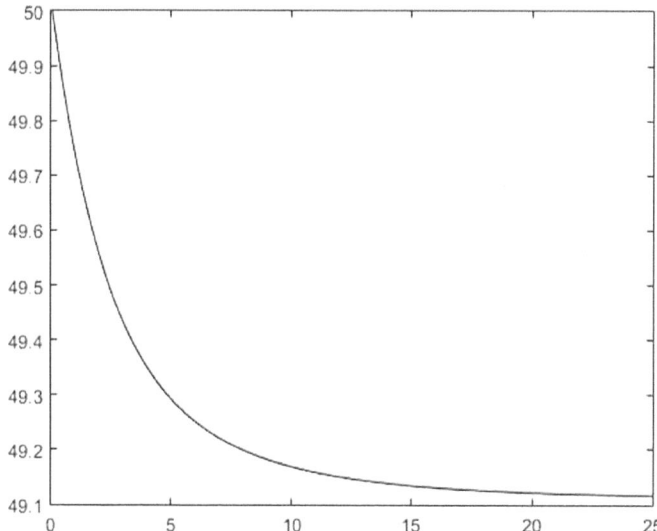

Fig. 7.7 Time variation of frequency of rendezvous

7.6   Conclusion

Frequency rendezvous dynamics is the dynamics for consensus frequency. It is different from dynamics of the angles that evolves from *initial* conditions (swing equation). FRD is a *chase* of various generator-frequencies to arrive at the frequency of rendezvous. Rendezvous frequency possesses dynamic fractals. FRD is relevant to high-inertia low-frequency dynamics that affects stability performance of the system. Multiple applications of FRD can be discovered in literature, namely coherency, dynamic power flows, dynamic voltages, ramping and many others. FRD may be useful in protection, too.

Reference

1. S.D. Varwandkar, M.V. Hariharan, Frequency excursions in complex power systems—a modular approach, in *International Conference on Power Systems (ICS 2017)*, Paper id. 046, 21–23 Dec. 2017, Pune, India

Chapter 8
MW-Blockchain

8.1 Introduction

Blockchain is a ledger which keeps systematic record of all transactions between various parties in the market. Same principle applies to MW-transactions from generator to bus to another bus, and finally to loads in power networks. A chain is thus formed by these transactions finally sinking into various loads. MW-transactions depend on network-impedance matrix. Fractals are used to compute MW-lineflows from the MWs of generators. When placed side-by-side, the chains of all generators form a blockchain. MW-Cost of electricity at user-end can be determined from the blockchain if MW-costs at the source are known.

8.2 Preliminaries

In deregulated power systems, schedules are decided by supplier-offers and demand-bids. Market price of electricity is result of a price-clearing mechanism. Suppliers who offer cheaper electricity get larger share of the market, limited only by their capacities [1]. Economic dispatch, per se, is no longer the principle for scheduling the generators. There is a core cost of generator associated with its MW-injection. MWs are dissipated in transmission lines and loads. A need thus arises to determine MW-costs associated with network losses and loads for computing the user-end costs [2, 3]. Transmission-service costs are regulated by transmission system operator and the market regulator. Costs for voltages are indirect and difficult to determine analytically [4]. Composition of costs must satisfy customers and the suppliers. It should lend itself to logical disbursal of revenue among various suppliers. Tracing is not very satisfactory but has been widely used for this purpose [5–7]. Fractal tomographs can be used more effectively. The overheads for voltages can also be computed using

© The Author(s), under exclusive license to Springer Nature Singapore Pte Ltd. 2023   75
S. D. Varwandkar and M. V. Hariharan, *Fractal Tomography for Power Grids*,
SpringerBriefs in Computational Intelligence,
https://doi.org/10.1007/978-981-99-3443-0_8

tomographs. Surging the prices, when operational limits are crossed, can be logically implemented by regulators. Disbursement of revenue among suppliers can be made logical and hence fair.

8.3 Blockchain

First step is to obtain fractal tomographs for lineflows and voltages. See also [8–10]. MWs are replaced by their respective costs to obtain the blockchain for costs. Congestion costs for excess MW-flow can be computed. Incentive and disincentives can be deviced.

Consider a system description,

$$
\underline{y} = \begin{bmatrix} [M] \\ [N] \end{bmatrix} \underline{x} \tag{8.1}
$$

Coefficient matrix in (8.1) is non-square with submatrix $[M]$ assumed invertible. $[N]$ is the null space. This model is called descriptor model. Such models are found in power systems [11]. Load impedances at generator buses are denoted by subscript Ldg and at non-generator buses by Ldx. A vector of load powers at generator buses, load powers at non-generator buses, and the lineflows, is written as a single vector,

$$
\begin{bmatrix} \underline{p_{Ldg}} \\ \underline{p_{Ldx}} \\ \underline{p_f} \end{bmatrix} = \begin{bmatrix} [\varepsilon_{Ldg}] \\ [\sigma_{Ldx}] \\ [\sigma_f] \end{bmatrix} \underline{P} \tag{8.2}
$$

$$
\varepsilon_{Ldg,i} = \operatorname{Re}\left(\frac{1}{R_{ii}} \xi_{ei} \xi_{ei}^* y_e^* \right) \text{ for } e \equiv Ldg
$$

$$
\sigma_{Ldx,i} = \operatorname{Re}\left(\frac{1}{R_{ii}} \xi_{ei} \xi_{ei}^* y_e^* \right) \text{ for } e \equiv Ldx
$$

$$
\sigma_{f,i} = \operatorname{Re}\left(\frac{1}{R_{ii}} \xi 1_{ei} \xi_{ei}^* y_e^* \right) \text{ for } e = \text{line-elements}
$$

Vectors of load-voltage deviations from unity $\left| \frac{v^2}{v_{base}^2} - 1 \right|$ and for line-loss $\underline{p_{LL}}$ in line elements can be conveniently added to (8.2) because both have same number of columns, so that,

$$
\begin{bmatrix}
\underline{p_{Ldg}} \\
\underline{p_{Ldx}} \\
\underline{p_f} \\
\underline{v^2} \\
\underline{p_{LL}}
\end{bmatrix}
=
\begin{bmatrix}
[\varepsilon_{Ldg}] \\
[\sigma_{Ldx}] \\
[\sigma_f] \\
[\eta] \\
[\beta]
\end{bmatrix}
\underline{P}
\tag{8.3}
$$

Rows in the five sub-vectors on the LHS are of dimensions *ng, nb-ng, ne-nd, nd* and *ne-nd*, respectively (*ng* = generator buses, *nb* = total buses, *ne* = number of 'transmission lines + loads' and *nd* = number of loads). Columns are *ng* in number for all sub-matrices in (8.3). Matrix $[\varepsilon_{Ldg}]$ is invertible. All sub-matrices other than $[\varepsilon_{Ldg}]$ fall within the sub-matrix $[N]$ of (8.1) and constitute the null space as per conventional matrix properties. Variables in $[N]$ are however real vectors $\underline{p_{Ldx}}$, $\underline{p_f}$, $\underline{v^2}$ and $\underline{p_{LL}}$. Costs of these relate to costs of generator-MWs at the source according to fractals of the respective tomographs. Transmission costs are prescribed by the regulator, for example, $k_f\%$ of the MW-flow on the line. Let the row elements of $\underline{CP}$ denote costs of MWs injected by the generators. The cost model is then given by (8.4).

$$
\begin{bmatrix}
\dfrac{C_{Ldg}}{C_{Ldx}} \\[2pt]
\dfrac{C_f}{} \\[2pt]
C_{\left(\frac{v^2}{v_{base}^2}-1\right)} \\[2pt]
\underline{C_{LL})}
\end{bmatrix}
=
\begin{bmatrix}
[\varepsilon_{Ldg}] \\
[\sigma_{Ldx}] \\
k_f * [\sigma_f] \\
[\eta] \\
[\beta]
\end{bmatrix}
\underline{CP}
\tag{8.4}
$$

MW (content)-costs at the user-ends are sum of rows of $\underline{C_{Ldg}}$, $\underline{C_{Ldx}}$ and $\underline{C_{LL}}$ of (8.4) and are given by (8.5). There are *ng* components for each "row-sum" corresponding to generators. Vertical column-wise terms (*ng* in number) are the costs payable by various loads to the respective generators (columns). Revenue can thus be precisely allocated in the formulation itself.

$$
\begin{bmatrix}
C_{Ldg1} \\
C_{Ldg2} \\
. \\
C_{Ldg,ng} \\
\hline
C_{Ldx,ng+1} \\
C_{Ldx,ng+2} \\
. \\
C_{Ldx,nb} \\
\hline
C_{L1} \\
C_{L2} \\
. \\
C_{Lnl}
\end{bmatrix}
=
\begin{bmatrix}
(\varepsilon_{Ldg1,1}P_1)C_1 + (\varepsilon_{Ldg1,2}P_2)C_2 + (\varepsilon_{Ldg1,3}P_3)C_3 + \cdots \\
(\varepsilon_{Ldg2,1}P_1)C_1 + (\varepsilon_{Ldg2,2}P_2)C_2 + (\varepsilon_{Ldg2,3}P_3)C_3 + \cdots \\
. \\
(\varepsilon_{Ldgng,1}P_1)C_1 + (\varepsilon_{Ldgng,2}P_2)C_2 + (\varepsilon_{Ldgng,3}P_3)C_3 + \cdots \\
\hline
(\sigma_{Ldx(ng+1),1}P_1)C_1 + (\sigma_{Ldx(ng+1),2}P_2)C_2 + (\sigma_{Ldx(ng+1),3}P_3)C_3 + \cdots \\
(\sigma_{Ldx(ng+2),1}P_1)C_1 + (\sigma_{Ldx(ng+2),2}P_2)C_2 + (\sigma_{Ldx(ng+2),3}P_3)C_3 + \cdots \\
. \\
(\sigma_{Ldxnb,1}P_1)C_1 + (\sigma_{Ldxnb,2}P_2)C_2 + (\sigma_{Ldxnb,3}P_3)C_3 + \cdots \\
\hline
(\beta_{L1,1}P_1)C_1 + (\beta_{L1,2}P_2)C_2 + (\beta_{L1,3}P_3)C_3 + \cdots \\
(\beta_{L1,1}P_1)C_1 + (\beta_{L1,2}P_2)C_2 + (\beta_{L1,3}P_3)C_3 + \cdots \\
. \\
(\beta_{Lnl,1}P_1)C_1 + (\beta_{Lnl,2}P_2)C_2 + (\beta_{Lnl,3}P_3)C_3 + \cdots
\end{bmatrix}
\tag{8.5}
$$

Lineflow fractals and voltage fractals are used for determining cost of line-flows and voltages. Voltage quality is indicated by deviation from nominal voltage. Blockchain of costs for lineflow service and voltage quality is included in the following blockchain.

$$
\begin{bmatrix}
C_{f1} \\
C_{f2} \\
. \\
C_{fnl} \\
\text{----} \\
C_{(v_1^2/v_{base}^2 - 1)} \\
C_{(v_2^2/v_{base}^2 - 1)} \\
. \\
C_{(v_{nb}^2/v_{base}^2 - 1)}
\end{bmatrix}
=
\begin{bmatrix}
(k_f * \sigma_{f1,1} P_1)C_1 + (k_f * \sigma_{f1,2} P_2)C_2 + (k_f * \sigma_{f1,3} P_3)C_3 + \cdots \\
(k_f * \sigma_{f2,1} P_1)C_1 + (k_f * \sigma_{f2,2} P_2)C_2 + (k_f * \sigma_{f2,3} P_3)C_3 + \cdots \\
. \\
(k_f * \sigma_{fnl,1} P_1)C_1 + (k_f * \sigma_{fnl,2} P_2)C_2 + (k_f * \sigma_{fnl,3} P_3)C_3 + \cdots \\
\text{--------------------} \\
(\eta_{1,1} P_1)C_1 + (\eta_{1,2} P_2)C_2 + (\eta_{1,3} P_3)C_3 + \cdots \\
(\eta_{2,1} P_1)C_1 + (\eta_{2,2} P_2)C_2 + (\eta_{2,3} P_3)C_3 + \cdots \\
. \\
(\eta_{nb,1} P_1)C_1 + (\eta_{nb,2} P_2)C_2 + (\eta_{nb,3} P_3)C_3 + \cdots
\end{bmatrix}
\tag{8.6}
$$

8.4 Examples

Consider a simple system shown in Fig. 8.1. Base MVA $= 100$. The network has all resistances in pu. Generator buses 1 and 2 have local loads of 1 and 1.5 MW (*not shown in the figure*). Subsystems for the generators are also shown. Generation costs are ₹3000 and ₹3200 per MW for the two generators.

Objective is to demonstrate the structure of the blockchain.

The cost model (8.4), for the base case in ₹, is given by,

Fig. 8.1 A simple example

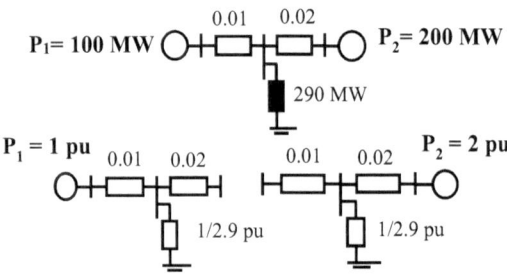

$$\begin{bmatrix} C_{Ldg} \\ C_{Ldx} \\ C_f \\ C_{\left(\left|\frac{v^2}{v_{base}^2}-1\right|\right)} \\ C_{LL)} \end{bmatrix} = k_f * \begin{bmatrix} \begin{bmatrix} 0.003518 & 0.003229 \\ 0.004979 & 0.005425 \end{bmatrix} \\ [0.96328 \quad 0.93664] \\ \begin{bmatrix} 0.99648 & -0.003229 \\ -0.004979 & 0.99457 \end{bmatrix} \\ \begin{bmatrix} 0.0 & 0.0 \\ 0.0 & 0.0 \end{bmatrix} \\ \begin{bmatrix} 0.02822 & \approx 0 \\ \approx 0 & 0.0570 \end{bmatrix} \end{bmatrix} \begin{bmatrix} 3000 * 100 \\ 3200 * 200 \end{bmatrix}$$

For the base case, there is no voltage charge. It applies only when voltage deviates from the base case. Constant k_f determines transmission charge/MW of lineflow. Price for voltage is calculated from,

$$C_{\left(\left|\frac{v_i^2}{v_{i\,base}^2}-1\right|\right)} = \left(\left|\frac{v_i^2}{v_{i\,base}^2}-1\right|\right) * C_i P_i$$

"Voltage" in our analysis means SD-voltage. The blockchain for MW-costs, (see (8.4)),

$$\begin{bmatrix} C_{Ldg} \\ C_{Ldx} \\ C_{LL)} \end{bmatrix} = ₹ \begin{bmatrix} \begin{bmatrix} 738.8 & + & 1446.7 \\ 1045 & + & 2430.4 \end{bmatrix} \\ [202290 + 419620] \\ \begin{bmatrix} 84.674 + 0.0020669 \\ 0.0044818 + 350.08 \end{bmatrix} \end{bmatrix}$$

Loading the system above the base case can be implemented by varying components of $\underline{P}$ vector in (8.4). Blockchain for transmission and voltage quality for the base case turns out to be,

$$\begin{bmatrix} C_f \\ C_{\left(\frac{v^2}{v_{base}^2}-1\right)} \end{bmatrix} = ₹ \begin{bmatrix} \begin{bmatrix} 10167 + 72.327 \\ 52.269 + 21053 \end{bmatrix} \\ \begin{bmatrix} 0 & + & 0 \\ 0 & + & 0 \\ 0 & + & 0 \end{bmatrix} \end{bmatrix}$$

Example 2 Six-bus system used by Garver [12] for network planning is shown in Fig. 8.2. Base MVA $= 100$. Following changes have been made in the original data: the decimal point for impedances of Lines 1–2, 1–4 and 2–4 are shifted to the left to make it comparable with other lines to make it realistic. In the original data, these line impedances are very high (probably a typo) which makes it a weak interconnection of three systems which are basically radial and hence not realistic. A small load of 30 MW was added at bus 6. Bus 6 is used as slack bus in the conventional Newton–Raphson method, but in our method slack bus is not needed. Data are shown in Tables 8.1, 8.2 and 8.3.

Fig. 8.2 Six-bus system
(Garver [12])

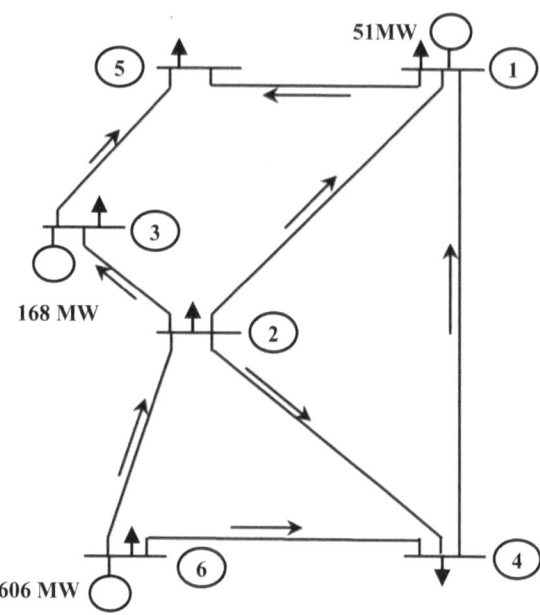

Table 8.1 Generator data

Generator	At bus	MW	₹/MWh
1	1	51	3000
2	3	168	3200
3	6	606	2500

Table 8.2 Line data, pu

Line	R	X	B
1–2	0.01	0.04	0.0001
1–4	0.015	0.06	0.0001
1–5	0.05	0.2	0.0001
2–3	0.05	0.2	0.0001
2–4	0.01	0.04	0.0001
2–6	0.0188	0.075	0.0001
3–5	0.025	0.1	0.0001
4–6	0.0375	0.15	0.0001

Load is increased in steps of 10%, starting from the base values given in Table 8.3. The user-end costs (UEC) are plotted in Fig. 8.3. In the bar-groups, the first bar denotes the base case, the second, 110% loading, the third 120%, and so on. There are six bar-groups for the six buses. Locational criticality of bus 3 is obvious from this plot. Computed costs are surged by 20% when UEC crosses ₹3200/MWh (Fig. 8.4).

Table 8.3 Load data

Load at bus	MW	MVAR
1	80	20
2	240	50
3	40	10
4	160	40
5	240	50
6	30	10

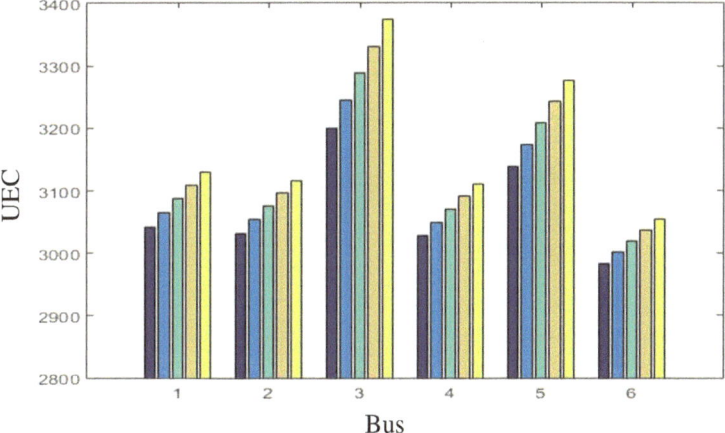

Fig. 8.3 User-end costs without surging ₹/MWh

At bus 3, surging applies to all load levels. However, at bus 5, it applies only from 130% loading onwards. A warning to reduce consumption can therefore be given to customers at bus 3 and bus 5.

8.5 Concluding Remarks

MW-blockchains for user-end costs can be constructed following the procedure described in this chapter. User-end electricity prices can be obtained by adding to it, service costs and voltage-quality costs. The latter serves as a substitute for reactive-power costs. Blockchain is useful as a check on pricing practices. It also has potential for new market designs.

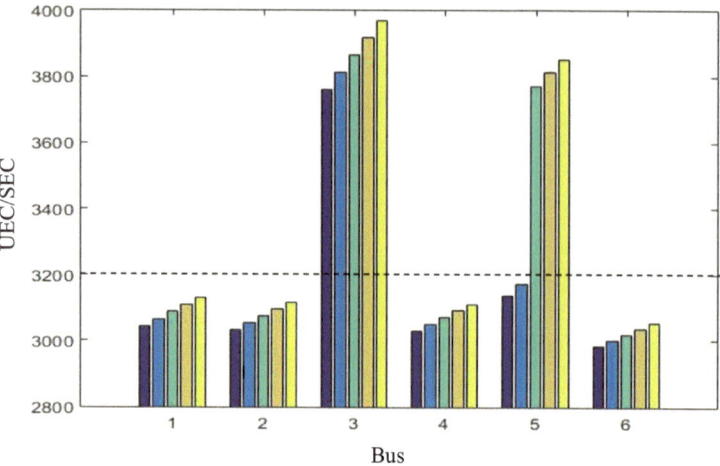

Fig. 8.4 Surging of user-end costs at violation buses, ₹/MWh

References

1. M.C. Caramanis, R.E. Bohn, F.C. Scheweppe, Optimal spot pricing: practice and theory. IEEE Trans. PAS **101**(9), 3234–3245 (1982)
2. D. Cheverez-Gonzalez, C.L. DeMarco, Admissible locational prices via Laplacian structure in network constraints. IEEE Trans. Power Syst. **24**, 125–133 (2009)
3. L. Chen, H. Suzuki, T. Wachi, Y. Shimura, Component of nodal prices for electric power systems. IEEE Trans. Power Syst. **17**(1), 41–49 (2002)
4. Y. Liu, F.F. Wu, Impacts of network constraints on electricity market equilibrium. IEEE Trans. Power Syst. **22**(1), 126–135 (2007)
5. J. Bialek, Tracing the flow of electricity. IEE Proc. Gener. Transm. Distrib. **143**(4), 313–320 (1996)
6. A.R. Abhyankar, S.A. Soman, S.A. Khaparde, Min-max fairness criteria for transmission fixed cost allocation. IEEE Trans. Power Syst. **22**(4), 2004–2104 (2010)
7. Y.C. Chen, S.V. Dhople, Tracing power with circuit theory. IEEE Trans. Smart Grid **11**(1), 138–147 (2020)
8. S.D. Varwandkar, J. Lin, Direct determination of load flow quantities from PQ injections: a new formulation. Int. J. Electr. Power Energ. Syst. **89**, 19–26 (2017)
9. S.D. Varwandkar, Realification of power flow. IEEE Trans. Power Syst. **34**(3), 2433–2440 (2019)
10. M.V. Hariharan, S.D. Varwandkar, P.P. Gupta, *Modular Load Flow for Restructured Power Systems, Lecture Notes in Electrical Engineering (LNEE, er. 374)* (Springer, 2016)
11. J. Lin, Md.R. Hesmazadeh, O. Galland, Application of null space method in computing electricity prices with voltage-stability constraints, in *North American Power Systems Conference (NAPS)* (2014)
12. L.L. Garver, Transmission network estimation using linear programming. IEEE Trans. Power Appl. Syst. **89**(7), 1688–1697 (1970)